Physiological Optical Imaging

Physiological optical imaging is a group of emerging technologies that aim to provide healthcare practitioners and biomedical researchers with information about tissue physiology or pathophysiology using approaches different from traditional medical imaging (PET, ultrasound, MRI, X-ray, or CT scan).

This book provides a comprehensive review of this group of technologies, combining a current medical literature review with an overview of cutting-edge technologies, the physics behind them, and common features across different technologies. It presents technical and physiological considerations that impact the sensitivity and spatial resolution of each technology and practical ways to improve them. The book emphasises low-cost technologies, which can be implemented in point-of-care settings. It is illustrated primarily with examples from wound care and oncology, with additional examples from other medical fields, including ophthalmology and neuroimaging.

It can be used as a one-stop reference and practical guide for healthcare professionals, clinical researchers, and engineers working with emerging bioimaging technologies who are looking to utilise physiological optical imaging technologies in biomedical research or clinical practice.

Key Features:

- Provides a comprehensive review of current technologies.

- Written as a practical guide with physiological and design considerations and illustrations.

- Presents a 360-degree view on the topic: a combination of clinical information alongside a technological background.

Gennadi Saiko is an accomplished scientist and entrepreneur with a successful track record in world-famous research institutes (Institute of General Physics, Moscow, Princess Margaret Hospital, Toronto), Fortune 500 companies (American Express), world-renowned universities (Friedrich-Alexander-University of Erlangen-Nürnberg, Moscow Institute of Physics and Technology, Toronto Metropolitan University), and startups (Infinium Capital, Oxilight Inc, Swift Medical Inc). Gennadi has received MSc and Ph.D. in physics from the Moscow Institute of Physics and Technology. His primary interest is in the development of innovative optical imaging modalities for healthcare, with primary focus on wound care, cardiology, emergency, and critical care. He combines both academic and industrial expertise. Gennadi has successful experience in IP development and commercialization in MedTech. In particular, he developed a handheld multimodal imaging system primarily used in wound care, which Swift Medical Inc commercialized as Swift Ray 1. Gennadi authored more than 40 publications in peer-reviewed journals, two books, seven book chapters, and several patents. In addition, he delivered more than 30 presentations at international scientific conferences and chaired sessions at multiple international conferences. Gennadi is a reviewer at more than 20 peer-reviewed journals and serves on boards of several scientific journals. Gennadi is Adjunct Professor of the Department of Physics at Toronto Metropolitan University (formerly Ryerson University).

Physiological Optical Imaging

Gennadi Saiko

CRC Press
Taylor & Francis Group
Boca Raton London New York

CRC Press is an imprint of the
Taylor & Francis Group, an **informa** business

Designed cover image: © Shutterstock

First edition published 2025
by CRC Press
2385 NW Executive Center Drive, Suite 320, Boca Raton FL 33431

and by CRC Press
4 Park Square, Milton Park, Abingdon, Oxon, OX14 4RN

CRC Press is an imprint of Taylor & Francis Group, LLC

ISBN: 978-1-032-76778-9 (hbk)
ISBN: 978-1-032-77330-8 (pbk)
ISBN: 978-1-003-48250-5 (ebk)

DOI: 10.1201/9781003482505

Typeset in Minion
by Apex CoVantage, LLC

Contents

Foreword

I F YOU SEARCH FOR "Physiological Imaging" at the time of this book's writing (end of 2023–beginning of 2024), Google returns results about physiologic or functional imaging by traditional medical imaging modalities and their variations: MR, PET, etc. Occasionally, it will mention fNIRS as an optical modality, even though it is not truly an imaging modality per se.

However, with advancements in inexpensive optical sensors, particularly CMOS, a rapidly growing range of novel imaging technologies provides invaluable clinical information to medical professionals. They include hyperspectral imaging, bacterial imaging, thermography, and OCT, to name a few. These technologies can also extract information about the physiology or pathophysiology of tissues. As such, these technologies can be termed "Physiologic Imaging" as well. Pertinent to optical modalities, I heard this term for the first time from Robert (Bob) Bartlett, MD, in early 2020. He used it to differentiate from traditional RGB imaging, which can be called "anatomical imaging."

Thus, to avoid confusion and make the term more descriptive, I titled the book *Physiological Optical Imaging*.

As these technologies are in a rapid development stage, most of them are developing in silos. However, there are a lot of commonalities between these approaches. This book is an attempt to find these commonalities and share best practices across different domains.

I wrote this book for a broad range of healthcare professionals, clinical researchers, engineers, engineering, and medical students who are dealing with developing new bioimaging technologies, using these technologies in biomedical research, and translating these technologies into clinical practice.

I am very thankful to Robert (Bob) Bartlett, MD, for numerous inspiring discussions, which helped me conceptualize the book. I am thankful to

Prof. Alexandre (Sasha) Douplik, Ph.D., my research partner, and numerous past and present graduate and undergraduate students from the Toronto Metropolitan University (formerly Ryerson University) and the University of Waterloo, who worked with me on numerous optical modalities for healthcare. Finally, I am very thankful to my wife, Vera, for her ongoing support and understanding, and my parents for their dedication.

CHAPTER 1

What Is Physiological Optical Imaging?

Gᴸᴏʙᴀʟ ʟɪꜰᴇ ᴇxᴘᴇᴄᴛᴀɴᴄʏ ʜᴀꜱ changed dramatically in the last century. While in 1900, the average life expectancy of a newborn was 32 years, by 2021, this had more than doubled to 71 years [1].

This remarkable increase can be attributed to advances in medicine, public health, and living standards.

However, in recent years, this progress has slowed down and flattened. In some cases, we even see its reversal, at least on a regional level. For example, Statistics Canada found that the life expectancy of Canadians fell to 81.3 years in 2022 from 81.6 years in 2021 [2]. What is even more worrisome is that life expectancy has declined for the third year in a row [2],

> While this immediate trend can be at least partially attributed to the COVID-19 pandemic, other factors are also at play. In particular, shifts in diet, lifestyle, and exercise drive the growth of obesity and associated health conditions in all countries.

For example, in 2021, 529 million people were living with diabetes worldwide [3]. By 2050, more than 1.31 billion people are projected to have diabetes, with expected age-standardized total diabetes prevalence rates greater than 10% in two super regions: 16.8% in North Africa and the Middle East and 11.3% in Latin America and the Caribbean [3].

DOI: 10.1201/9781003482505-1

These numbers can be significantly higher for certain age and ethnical groups. For example, in 2013, 37.3 million Americans, or 11.3% of the population, had diabetes. For senior citizens older than 65 years old, the prevalence was 27%. The United States Center for Disease Control estimates for prediabetes is 38%, which continues to rise annually.

Similarly, there is a substantial growth in the prevalence of cardiovascular disease (CVD), which remains the leading cause of disease burden in the world. According to the Global Burden of Diseases 2019 (GDP 2019) study, prevalent cases of total CVD nearly doubled globally from 271 million in 1990 to 523 million in 2019, and the number of CVD deaths steadily increased from 12.1 million in 1990, reaching 18.6 million in 2019 [4]. In addition to aging and population growth, certain risk factors that are on the rise in prevalence further drive this growth, including high systolic blood pressure, high fasting plasma glucose, high low-density lipoprotein cholesterol, high BMI, dietary risks, and low physical activity.

Similarly, according to the International Agency for Research on Cancer, the global cancer burden rose to 19.3 million cases and 10 million deaths in 2020 and is predicted to increase to more than 30 million cases and over 16 million deaths by 2040, with most of the increases occurring in low- and middle-income countries [5].

Finally, the COVID-19 pandemic raised the prominence of infectious diseases as a global threat. Infectious diseases were always a threat. For example, the Black Death or bubonic plague was one of humankind's worst pandemics that killed more than 25 million people or at least one-third of Europe's population during the fourteenth century. However, with recent advances in medicine and public health, many diseases have been eradicated or significantly mitigated. Thus, this threat was substantially subdued in the pre-COVID world, particularly in public opinion. However, the recent COVID-19 pandemic demonstrated vulnerabilities in the health systems and society.

These new threats and challenges require a novel, comprehensive, multifactorial approach. While new effective treatments are being developed, moving diagnostics to an earlier time point becomes increasingly important, as earlier diagnostics are associated with better outcomes and lower overall costs to healthcare. However, traditional medical imaging technologies (ultrasound, X-ray, CT scan, MRI, and PET) cannot be of great help here. While being instrumental in any aspects of clinical diagnostics, they are confined to specialized laboratories. They are expensive (with ultrasound as an exception), labor intensive, and require expensive equipment and specialized training. As such, they will still (at least in the foreseeable future) be a) confined to specialized laboratories, which limits their utility in early-stage diagnostics, and b) not universally available across different geographies. Thus, a paradigm change may be required. In particular, novel, simpler diagnostic techniques are necessary. This is where optical technologies may play an essential role.

Optical technologies are abundant in medicine. They are used across the whole clinical workflow. For example, a pulse oximeter is probably the most ubiquitous and well-known optical medical device.

This book is about optical imaging technologies, which are used or can be used in the medical field to extract information about tissue composition or physiological processes. We use the term "physiological imaging" to differentiate this group of technologies from "anatomical imaging." What do we mean by this differentiation?

Anatomical imaging refers to traditional RGB or monochrome imaging. Anatomical imaging is the most ubiquitous imaging modality in medicine, as multiple images are taken regularly for documentation purposes across different levels of care. For example, in a typical anatomical imaging scenario in wound care, the nurse takes an image of the wound using a DSLR camera or a smartphone. This image is then either stored in an electronic health record, EHR, or processed, and some geometrical information (e.g., wound area, length, width, depth) is retrieved automatically, semiautomatically, or manually. Then, this information can be used to track the healing progress in time. Thus, anatomical imaging is particularly helpful for monitoring, including the measurement and documentation of wounds. However, despite being an invaluable medical technology, anatomical imaging is not the focus of this book as it does not provide an explicit screening, triaging, or diagnostic functionality beyond "naked eye" examination.

While anatomical imaging is probably the first idea that comes to mind when you think about optical imaging in healthcare, there is a surprisingly rich pool of optical imaging technologies that are used or can potentially be used to extract physiological or pathophysiological parameters. We refer to them as physiological imaging. To differentiate them from physiological imaging derived from traditional medical imaging, we use the modifier "optical." Thus, the descriptive term will be "physiological optical imaging."

In physiological optical imaging, the imaging technology is augmented by some extra capability, such as laser Doppler, hyperspectral, or fluorescence imaging, to provide information beyond "naked eye" examination and is used for screening, triaging, and diagnostic purposes. This group of technologies is the primary focus of this book. In particular, current optical diagnostic techniques and physiological imaging will be considered in Part 2 (Large-Area Physiological Optical Imaging Techniques) and Part 3 (Endoscopy and Microscopy).

However, the anatomical and physiological imaging border is not well defined. For example, consider taking a color picture of the skin area. In this case, if the picture is taken for documentation purposes or to measure the wound sizes (length, width, area, etc.), it is clearly anatomical imaging. However, if the picture is taken to assess whether the erythema is present, we will consider this case a physiological imaging.

A few words about what we mean by "optical." The word "optics" comes from the ancient Greek word ὀπτική (*optikē*), meaning "appearance, look." Thus, the narrow meaning of "optics" is related to visible light. However, in physics, optics usually describes the behavior of visible, ultraviolet (UV), and infrared light.

What makes visible, ultraviolet (UV), and infrared light unique? While light is an electromagnetic wave and shares common features with other types of radiation, such as X-rays or radio waves, we can define the optical range based on two common features that are broadly used in optics and optical instrumentation: a) extensive use of refraction (geometrical optics) and diffraction (physical optics) of light and b) detection using the quantum (photoelectric or photoconductive) effects. These two features allow for more precise short- and long-wave limit definitions of the optical range.

For example, the shortwave limit can be defined as the boundary between UV light and X-rays, as most X-ray techniques do not rely on refraction or diffraction (other than in crystallography and material science) of light. We can define this shortwave limit even more strictly as "nonionizing" radiation. This restriction limits photon energy to 10eV or 121.6 nm.[1]

Similarly, the long-wave limit can be drawn between terahertz and millimeter waves. While the former can be detected in numerous ways, including photoconductive effect and antennas, the latter can be detected using only antennas. Moreover, diffraction optics is well established in the terahertz range.

Based on this definition, in this book, we aim to cover the extended optical spectral range, from UV to visible to all shades of infrared to terahertz spectrum ranges.

This book focuses on imaging techniques, as they provide additional spatial information about underlying tissue and its physiology. As such, we will exclude explicit single-point measurement techniques. For example, we will consider thermographic imaging in thermography and ignore single-point temperature measurements, like routine temperature checks using a mercury thermometer. Similarly, we will exclude traditional NIRS techniques (other than fNIRS), contact photoplethysmography, and pulse oximetry. However, there is a multitude of imaging modalities that stem from them (e.g., perfusion imaging). These modalities will be considered in Part 2.

As light penetration depth in tissues is typically limited to millimeters, most physiological imaging modalities are limited to superficial tissues: skin and mucosa (with the notable exceptions of fNIRS and diffuse optical tomography).

As the number of pathologies, even for superficial tissues, is vast, we will illustrate the clinical utility of physiological optical imaging technologies with a primary focus on two large groups of pathologies with high incidence and significant impact on patients' lives and healthcare in general: wounds and carcinomas. However, in some cases, we will illustrate the utility for other medical fields. Why did we pick these two pathologies? Because these pathologies have several common motifs. In particular, they originate from the tissue surface or close to the tissue surface (e.g., basal carcinoma). That is why optical modalities, with their small penetration depth, are uniquely well positioned to investigate them. Moreover, a broad range of technologies (virtually all we discuss in this book) can be used for their diagnostics. Thus, to some extent, we can compare "apple to apple" and see some commonalities, pros, and cons.

The primary aim of this book is to elucidate these commonalities. To achieve this, we can start with the most common feature of physiological optical imaging technologies: their small penetration depth. Depending on the technology, it ranges from tens or hundreds of micrometers to several millimeters. Thus, increasing sampling depth is the common motif across all these technologies. Two other common motifs are to increase resolution and imaging contrast. These three themes (sampling depth, resolution, and imaging contrast) are the framework of this book. We will consider their definitions and common approaches to improve them in Part 1 of the book. In Parts 2 and 3, we will consider different physiological optical imaging modalities focusing on these three factors.

At first glance, physiological or diagnostical imaging applies only to imaging a patient. However, it is a broader concept, which includes *ex vivo* and even *in vitro* studies. Thus, it is important to discuss the clinical settings where physiological imaging can provide significant value. We can consider two primary settings: point of care (POC) and clinical laboratories, where *in vitro* diagnostics (IVD) are used.

Clinical laboratory settings are easier to define, so we will start from this setting. Traditional laboratory testing typically involves a multiple-step process that includes collecting samples from the patient at the bedside or the clinic, transporting them to a centralized laboratory (often located far away), and then subjecting them to several often quite sophisticated and labor-intensive processing steps. The physiological imaging technologies can (and some of them do already) significantly decrease time and increase the throughput of laboratory testing. For example, they can automate histology reading or create high-throughput immunoassays or high-throughput screening (HTS).

However, as traditional laboratory testing is a time-consuming multistep process, the delay in treatment caused by these inefficiencies can hinder timely clinical decision-making. That is why moving clinical decision-making to point of care is of great importance. POC encompasses a broad range of settings, from physician offices to public health clinics to operating rooms (OR). They can also include pharmacies, ambulances, nursing homes, and even patient homes.

While POC modality typically refers to a modality used *in vivo* on a patient, a closely related group of testing technologies is used (or can be used) in POC settings. In particular, point-of-care testing (POCT) is clinical laboratory testing conducted close to the site of patient care where care or treatment is provided [6]. POCT provides rapid turnaround of test results with the potential to generate a result quickly so that appropriate treatment can be implemented, leading to improved clinical or economic outcomes compared to laboratory testing [7]. Thus, the POCT approach tries to tackle the inefficiencies of traditional laboratory testing.

While under POCT people typically understand testing strips, lateral-flow testing, immunoassays, antigen-based testing, and certain molecular testing [8], this concept can be extended to other applications like on-spot *ex vivo* cancer margin delineation.

Even though it can be perceived that virtually any clinical laboratory methodology can be moved closer to POC, this is incorrect. One of the critical characteristics of POCT is that it is easy to use [9]. Actually, it is an essential regulatory requirement that permits the use of clinical testing technology in POC settings. For example, in the US, the testing technology can be used in a POC setting if it has a CLIA Certificate of Waiver.[2] This requirement imposes significant limitations, for example, on sample preparation. As a result, POCT technologies should include minimal or no sample preparation at all. And this is where physiological imaging technologies shine. For example, certain technologies discussed later in the book (OCT, fluorescence imaging, vibrational imaging, terahertz imaging, to name a few) can detect cancer cells in freshly excised tissues close to real time, thus significantly reducing time for certain procedures, like Mohs surgery. Moreover, as they don't require sample preparation, the sample can further undergo standard clinical testing (e.g., staining).

These two settings (clinical laboratory and POC) and three use scenarios (clinical laboratory testing, POC, and POCT) will be the focus of this book, with the POC setting as the primary focus.

Also, we should briefly discuss the role of physiological optical imaging in the current healthcare model. It is unlikely that physiological optical imaging will replace traditional medical imaging technologies (ultrasound, X-ray, CT scan, MRI, and PET), at least in the foreseeable future. The medical field is a very conservative area, and very strong arguments are needed to be in place before changing any standard of care. Instead, physiological optical imaging can be considered complementary to traditional medical imaging. They can be used closer to the patient (e.g., in POC settings) for screening and triaging purposes. They can only be considered a primary diagnostic tool in certain scenarios (like the absence of traditional medical imaging).

The book is organized as follows.

Part 1 contains information about physical and physiological parameters, which define the capabilities of the imaging technologies and certain ways to overcome these limitations. These parameters include resolution, imaging contrast, and sampling depth.

Chapter 2 will overview resolution, which impacts the capabilities of optical modalities, and common approaches to increase it.

Chapter 3 will overview imaging contrast relevant to optical modalities and common approaches to increasing it.

Chapter 4 will overview the sampling depth and common approaches to increase it.

In Parts 2 and 3, we will overview existing optical diagnostic methods. We will focus on physiological imaging techniques, which have been translated into the clinic or used in clinical research. These methods focus on extracting physiological parameters used primarily for screening and diagnostic purposes. Efficient screening modalities allow earlier interventions in a patient population that presents clinically with late-stage complications, significant morbidity, and mortality risk.

We split the physiological optical imaging techniques into three groups: large-area imaging, endoscopic, and microscopic techniques.

In Part 2, we will consider large-area physiological optical imaging techniques. We further divide them based on the underlying physical principle into absorption-based (Chapter 5), scattering-based (Chapter 6), fluorescence-based (Chapter 7), thermography (Chapter 8), vibrational and terahertz (Chapter 9), and other technologies (Chapter 10). Nowadays, most diagnostic optical technologies used are imaging based. Imaging modalities have some inherited advantages that are particularly relevant for healthcare. Firstly, in most cases, they are noncontact (with ultrasound-based technologies, which require contact with skin, as a notable exception). It is beneficial to prevent bacterial contamination. It also benefits a broad range of patients (including people with diabetes) with sensitive skin. Secondly, it allows for extracting information from some areas, which allows immediate comparison between different skin areas (self-control). While some contact optical modalities (like photoplethysmographic or PPG devices) may still exist in labs, most incoming technologies are imaging based. Thus, the focus of this book will be primarily on imaging technologies.

The primary focus of Part 2 is large-area imaging modalities, which can be used for screening and triaging in large body areas (several to hundreds of sq. cm). However, in addition to that, there are multiple microscopic and endoscopic techniques. They will be considered in Part 3. In particular, Chapters 11 and 12 will cover endoscopic and microscopic techniques, respectively. Note that most techniques described in Part 2 can be (or already are) adapted for endoscopic and microscopic use. Thus, in Chapters 11 and 12, we will focus on distinct technologies used only in endoscopic or microscopic applications.

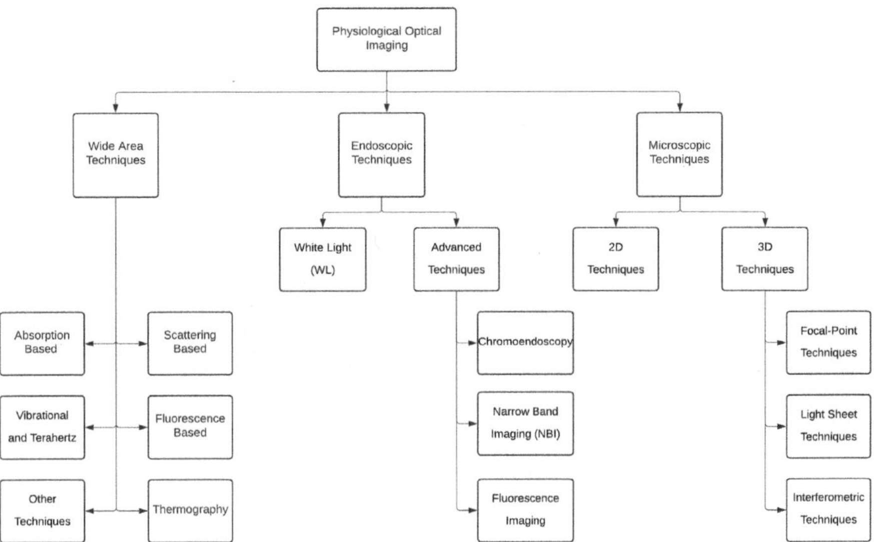

FIGURE 1.1 The proposed nomenclature of physiological optical imaging modalities.

Finally, in Chapter 13, we will discuss new frontiers in physiological imaging. With the rapid penetration of AI in all aspects of daily life, we cannot ignore its influence on medicine. Thus, this chapter will focus on applications of AI and other trends in medicine.

We tried to make the book self-contained and included as much supplementary information as possible. We put background information at the beginning of each chapter and moved supplemental information from the primary flow to avoid constantly switching the topic from physics and basic science to clinical utility. As such, we created Appendix A, where we will provide basic information on the morphology of superficial tissues (skin and mucosa) and certain pathologies (wounds and carcinomas). As all optical imaging modalities are based on light propagation in tissues, in Appendix B, we listed several models that can help us understand light propagation in tissues. They were used to derive certain results presented in the book.

Full disclosure: This book does not attempt to cover all the technologies. Any of the considered technologies and their clinical applications are very broad topics, probably worth writing a separate book. Instead, it attempts to briefly overview these technologies through a common framework of sampling depth, imaging contrast, and resolution with an identification of their pros and cons.

NOTES

1 Ultraviolet radiation with a wavelength shorter than 121.6 nm (hydrogen Lyman–alpha line) is called extreme ultraviolet (E-UV) and is considered ionizing. For example, US FCC defines 10eV (124nm) as the threshold for ionizing radiation.
2 As defined by FDA's CLIA, waived tests are categorized as "simple laboratory examinations and procedures that have an insignificant risk of an erroneous result."

REFERENCES

1. R Zijdeman, F Ribeira da Silva. "Life expectancy at birth (total)". (2015). https://hdl.handle.net/10622/LKYT53, IISH Data Collection, V1.
2. Death. "Statistics Canada". (2022). https://www150.statcan.gc.ca/n1/daily-quotidien/231127/dq231127b-eng.htm (accessed November 30, 2023)
3. GBD 2021 Diabetes Collaborators. "Global, regional, and national burden of diabetes from 1990 to 2021, with projections of prevalence to 2050: A systematic analysis for the Global Burden of disease study 2021". *Lancet* 402: 203–234 (2023).
4. GA Roth, GA Mensah, CO Johnson, et al. "Global Burden of cardiovascular diseases and risk factors, 1990–2019: Update from the GBD 2019 study". *J Am Coll Cardiol* 76(25): 2982–3021 (2020).
5. UICC: GLOBOCAN 2020: New Global Cancer Data. Released December 17, 2020. www.uicc.org/news/globocan-2020-new-global-cancer-data (accessed November 30, 2023).
6. CP Price. "Point of care testing". *BMJ* 322(7297): 1285–1288 (2001).
7. JH Nichols. "Utilizing point-of-care testing to optimize patient care". *EJIFCC* 32(2): 140–144 (2021).
8. MC Larkins, A Thombare. "Point-of-care testing". In *StatPearls [Internet]*. Treasure Island (FL): StatPearls Publishing (January 2023). www.ncbi.nlm.nih.gov/books/NBK592387/ (updated May 29, 2023)
9. A St John, CP Price. "Existing and emerging technologies for point-of-care testing". *Clin Biochem Rev* 35(3): 155–167 (2014).

PART 1

Factors That Impact Physiological Image

Typically, image quality is characterized by imaging contrast and resolution. However, as the physiological optical imaging looks into the tissue's depth, we will also be interested in sampling depth.

Thus, in this part of the book, we will consider basics across three dimensions: imaging contrast, resolution, and sampling depth.

It should be noted that these concepts are not independent but rather related. For example, sampling depth may impact the contrast. Similarly, the resolution will impact the contrast. However, as multiple researchers try to optimize them separately, we will consider them separately, too.

DOI: 10.1201/9781003482505-2

CHAPTER 2

Resolution of Physiological Optical Imaging Techniques

RESOLUTION IS AN IMPORTANT characteristic of any imaging system. Moreover, it is critically important for many physiological optical imaging applications, particularly microscopic techniques.

Common sense tells us that the higher the resolution, the better. However, the highest possible resolution is not always the answer, as it drives the cost and complexity of the system. Ideally, resolution needs to be optimized for a particular task. This chapter will focus primarily on technical considerations that drive resolution. However, physiological considerations should also be taken into account in physiological optical imaging. As this book focuses on imaging of superficial tissues like skin and mucosa, we will consider characteristic dimensions relevant to microcirculation in this chapter.

After analyzing technical and physiological determinants, we will consider safety determinants relevant to resolution and common ways to improve resolution.

2.1 TECHNICAL CONSIDERATIONS

Technical considerations refer to the factors impacted by the optical imaging system design. They depend on the selected optical elements, detector, and light wavelength(s) used in the particular imaging system.

DOI: 10.1201/9781003482505-3

2.1.1 Diffraction Limit

The resolving power of the imaging system is defined by the smallest distance that the lens can resolve. There are several interrelated approaches to determine it.

2.1.1.1 Abbe Diffraction Limit

The resolving power of the imaging system, δ, is connected to the light wavelength, λ and numerical aperture, NA of the objective lens:

$$\delta = \frac{\lambda}{2NA} \tag{2.1}$$

This expression is called the Abbe diffraction limit. The objective numerical aperture, NA, depends on the refractive index of the media between the objective lens and the specimen, n, and the half of the acceptance angle of the objective lens, a:

$$NA = n\sin(a) \tag{2.2}$$

Thus, the imaging system's resolution depends significantly on the imaging geometry, particularly the acceptance angle.

For example, for microscopic techniques, which can achieve the half acceptance angle, α, close to $\pi/2$, the primary limiting factor is the refractive index, as the objective numerical aperture, NA, cannot be more than the refractive index of the media between the objective lens and the specimen, n ($NA = n\,sin(a) < n$). Thus, for imaging in the air $(n = 1)$, the resolution of the microscopic technique can be on the scale of $\delta = \lambda/2$. For example, for 550 nm wavelength (green light), the resolution of the microscope can be on the scale of 275 nm, which is small compared to most biological cells (1 µm to 100 µm) but large compared to viruses (100 nm), proteins (10 nm) and less complex molecules (1 nm).

However, the primary limiting factor for large-area imaging techniques is the acceptance angle, which is typically small in these cases. For example, for 550 nm wavelength (green light), for the imaging system with a lens with a radius of 5 mm located 15 cm from the target area, the smallest resolved distance will be 8.25 µm.

Another approach to characterize the imaging system is to use the theoretical maximum resolving power of the lens given in line pairs per millimeter [lp/mm]. This measure of the diffraction-limited resolution, often referred to as the cutoff frequency of a lens, is calculated using the f-number of the lens ($f/\# = f/D$, where f is the focal length and D is the diameter of the aperture) and the wavelength of light, λ: [1]

$$\zeta = \frac{1}{f/\# \lambda} \tag{2.3}$$

For example, most smartphone cameras have an aperture between f/16 and f/20. [2]

Thus, for example, 550 nm wavelength (5.5×10^{-4} mm) results in 91–114 lp/mm, corresponding to 10 μm interline distance.

2.1.1.2 Point-Spread Function and Airy Disk

The resolution of an optical system can be characterized by a point-spread function (PSF), which describes the response of an optical imaging system to a point source or point object.

It should be noted that the magnification of the objective lens does not affect the PSF; it depends only on the NA and wavelength.

Diffraction on the aperture causes the divergence of the light or spread. In the case of the circular aperture, the PSF is characterized by a bright central disk (called the Airy disk), which, together with the series of concentric rings, is known as the Airy pattern.

As optical systems typically have a circular aperture, the size of the smallest feature in an image is limited by the size of the Airy disk.

Far from the aperture, the first null of the Airy pattern can be approximated as

$$\sin(\theta) \approx 1.22\lambda / D \tag{2.4}$$

Eq.2.4 is known as the Rayleigh criterion. Here θ is the angle at which the first minimum occurs, and D is the diameter of the aperture.

Similarly to Eq.2.3, we can link the radius of the Airy disk, r_{airy} with the f-number of the lens ($f/\#$) and numeric aperture of the lens (NA):

$$r_{airy} \approx 1.22\lambda f/\# \approx 0.61\lambda / NA \tag{2.5}$$

Note that in the air, the f-number of the lens (f/#) can be expressed through its numeric aperture (NA) as $f/\# = \dfrac{1}{2NA}$

2.1.2 Nyquist Criteria

Sampling refers to a process of converting a signal (for example, a function of continuous time or space) into a sequence of values (a function of discrete time or space).

The Shannon–Nyquist sampling theorem is a fundamental principle in digital signal processing that establishes a sufficient condition for a sample rate that permits a discrete sequence of samples to capture all the information from a continuous signal of a finite frequency range (bandwidth).

The original version of Shannon's theorem stated [3], "If a function x(t) contains no frequencies higher than B hertz, then it can be completely determined from its ordinates at a sequence of points spaced less than 1/(2B) seconds apart."

While the theorem was formulated for a single time-dependent variable, it can be extended to functions of arbitrarily many variables. As such, the theorem is applied broadly across temporal and spatial domains in various branches of science and engineering. In particular, it is broadly used in digital image processing to determine the minimum sampling rate required to avoid aliasing and to select band-limiting filters to keep aliasing below an acceptable amount when an analog signal is sampled.

In biomedical optics, the Shannon–Nyquist sampling theorem is applicable in both temporal and spatial domains.

The characteristic frequencies of physiological processes in physiological optical imaging are typically relatively low. For example, heart rate, respiratory rate, metabolic, and neurogenic oscillations are in the range of 0.02–3 Hz.

Thus, temporal sampling frequency is typically not a limiting factor for multidetector arrays like FPAs. However, it can be an issue for low-speed single-sensor acquisition systems like FTIR.

In the spatial domain, we can illustrate the application of the Shannon–Nyquist sampling theorem in a specific example. One particular application of physiological imaging is determining the parameters of the capillary grid, which can be used for shock progression monitoring [4] or cancer transformation detection.

In most skin parts, the capillaries are arranged vertically (hairpins). Thus, from the surface, they can be perceived as dots. However, the contrast between these points is relatively low. Therefore, to detect them, we have to account for two factors: a) the contrast ratio associated with them needs to be detectable by the sensor (pixel size ↑), and b) the spatial sampling frequency should satisfy the Shannon–Nyquist sampling theorem (pixel size ↓). In particular, the pixel size (and interpixel distance, respectively) should be more than 2x smaller than the typical intercapillary distance, which can be estimated from the capillary density. Some authors recommend sampling at least 2.3x the highest frequency.

2.1.3 Resolution of Different Imaging Technologies

In a typical scenario, the optical system forms an image of the target area at a lens' focal plane, where the image is sampled by a photodetector (e.g., scanning geometries) or an array of photodetectors (true imaging geometries). Thus, for homogeneous illumination (e.g., broad area illumination), the resolution of the system will be the size of the target area, which is projected on a single photodetector (e.g., pixel or cell).[1] For inhomogeneous illumination (e.g., laser beam in laser scanning technologies), the resolution of the system may be impacted by the size of the illuminated area (e.g., laser spot).

However, particular technologies may have other limiting factors. This section will consider several of the most commonly used signal acquisition geometries: "true" imaging using a multidetector array and laser scanning. In addition, we will consider the resolution of confocal microscopy and OCT as they have certain peculiarities.

2.1.3.1 Cameras

The most commonly used signal acquisition geometry in optical imaging is based on multisensory arrays, like CMOS. For such imaging geometry, the array of photosensors is often called a "focal plane array" or FPA. This term is mainly used in the infrared range of the spectrum. Other terms are used in specific spectral ranges. For example, in the visible spectrum range, an explicit name of the technology, like CMOS image sensor or CCD (charge-coupled device), is used more commonly than FPA.

Typically, FPA is a rectangular array of light-sensing cells or pixels. Thus, the imaging system's resolution is the field of view (FOV) of the optical system divided by the linear size of the FPA. The size of the pixel and the number of pixels depend on the spectral range. Whereas visible imagers such as CCD and CMOS image sensors are fabricated from silicon, using mature and well-understood processes, sensors in other spectral ranges must be fabricated from other, more exotic materials, which drives their cost. Arrays larger than 1024 × 1024 are available and commonly used in visible and infrared ranges of spectra. Much smaller arrays are available for other spectral ranges.

CMOS sensors with pixels smaller than 1.7 × 1.7 μm are used in consumer-grade cameras. However, they are characterized by strong noise. Machine vision cameras (C-mount) with resolutions from VGA to 2 megapixels usually have 4.6 to 6.5 μm pixels with 10–15 times larger light-active surfaces and much better signal-to-noise ratios.[5] However, much larger pixel sizes are used for low-light applications and small exposure times (on a scale of microseconds). For example, pixels with an edge length of 10–14 μm are commonly used in line scan cameras, given their short exposure time.

2.1.3.1.1 Interplay with Diffraction Limits Based on the pixel size and the selected lens, the imaging system may operate in three different regimes, as follows:

1. The diffraction-limited regime: if the pixel size is small compared to the spread of the diffraction PSF.

2. The instrument-limited regime: if the spread of the diffraction PSF is small compared to the pixel size.

3. The mixed regime, where both diffraction and instrumentation impact the resolution of the system, if the spread of the PSF and pixel size are similar.

For f/8 and 550 nm wavelength (green light), d_{airy} = 10.7 μm, similar to the pixel size for most commercially available "full frame" (43 mm sensor diagonal) cameras. Thus, these cameras will operate in the mixed regime for f-numbers around 8.

Cameras with smaller sensors typically have smaller pixels. However, their lenses can be designed for use at smaller f-numbers to operate in the mixed regime, which is generally optimal as the camera capabilities match those of the optical system. However, other regimes can be effective in certain circumstances.

2.1.3.2 Laser Scanning

Laser scanning is a technology used in many applications across various domains. For example, it is widely used for 3D object reconstruction (3D laser scanner).

Most laser scanners use moveable mirrors to steer the laser beam. While some applications (like barcode scanning) require just one-dimensional scans, most biomedical applications are interested in area scans, which require two-dimensional beam steering. To position a laser beam in two dimensions, it is possible to rotate one mirror along two axes (which is used mainly for slow scanning systems) or to reflect the laser beam onto two closely spaced mirrors mounted on orthogonal axes and driven separately.

The resolution of the laser scanner is driven by two factors: the smallest possible increment of the angle between two successive points and the size of the laser spot itself on the object.

Confocal microscopy is typically based on the laser scanning technology. As such, it is often called laser scanning confocal microscopy (LSCM). However, other factors also impact its resolution (see section 2.1.3.4. for details).

2.1.3.3 Wide-Field vs. Confocal Microscopy

Imaging techniques are characterized by the depth of field (DOF), the distance between the farthest and closest points in focus. For high NA geometry (e.g., microscopy), the depth of field can be approximated as

$$DOF \approx \frac{\lambda n}{NA^2} \tag{2.6}$$

Here, n is the refractive index of media between the sample and the lens.

However, in conventional (wide-field) microscopy, the objective collects light outside the focal area, decreasing contrast. The distinct feature of the confocal microscopy is the presence of a pinhole. The pinhole acts as a spatial filter by removing all emissions that do not originate from the focal plane. As such, the sample can be sliced or sectioned along the vertical axis. As a result of this design, the resolution of a confocal microscopy system is determined by the size and shape of the spot and the diameter of the pinhole [6]. Conventional wide-field microscopy can be considered a confocal microscopy with a very large pinhole.

For the wide-field imaging system, the lateral resolution (the XY plane) is determined by the distance from the center of the Airy disk to the first dark ring, r_{airy} (see Eq.2.5). For a confocal system, the pinhole radius is set typically smaller than r_{airy}, and thus the lateral resolution is [6]

$$\delta_{XY,conf} \approx \frac{0.8\lambda}{2NA} \tag{2.7}$$

Which is slightly smaller than the Airy radius.

The axial resolution (Z direction) is related to the depth of field (Eq.2.6). The axial resolution is determined by the distance from the center of the spot to the edge of the first Airy minimum in the vertical direction. Specifically, the axial resolution is defined by convention as one quarter of the distance between the first minima, above and below focus, along the axis of the three-dimensional diffraction image produced by the objective [7]. In addition to the wavelength λ and numeric aperture of the lens, NA, the axial resolution also depends on the refractive index of the media between the sample and the lens, n.

For wide-field imaging, the axial resolution is

$$\delta_{Z,wf} \approx \frac{2\lambda n}{NA^2} \tag{2.8}$$

For confocal imaging, the axial resolution is [6]

$$\delta_{Z,conf} \approx \frac{1.4\lambda n}{NA^2} \tag{2.9}$$

The pinhole size also impacts the resolution. Lateral resolution is more sensitive to pinhole size than axial resolution. It rapidly worsens as the pinhole is larger than one Airy unit or AU (the diameter of the Airy disk). A practical setting for confocal microscopy is a diameter of 1 AU. The dependence of axial and lateral resolutions on the pinhole size is depicted in Figure 2.1.

2.1.3.4 OCT

As OCT is an interferometric technique, the axial and lateral properties are decoupled from each other [9]. While lateral resolution is defined by the objective and the media in front of the sample, all axial properties are defined by the light source's coherence properties and the signal's sampling at the detector.

As OCT uses laser light sources with Gaussian beam profiles, the expressions for lateral resolution contain some additional numeric terms (including 2ln2, which accounts for the beam diameter at half maximum).

$$\delta_{XY,OCT} = \frac{\sqrt{2ln2}}{\pi} \frac{\lambda_0}{NA} \tag{2.10}$$

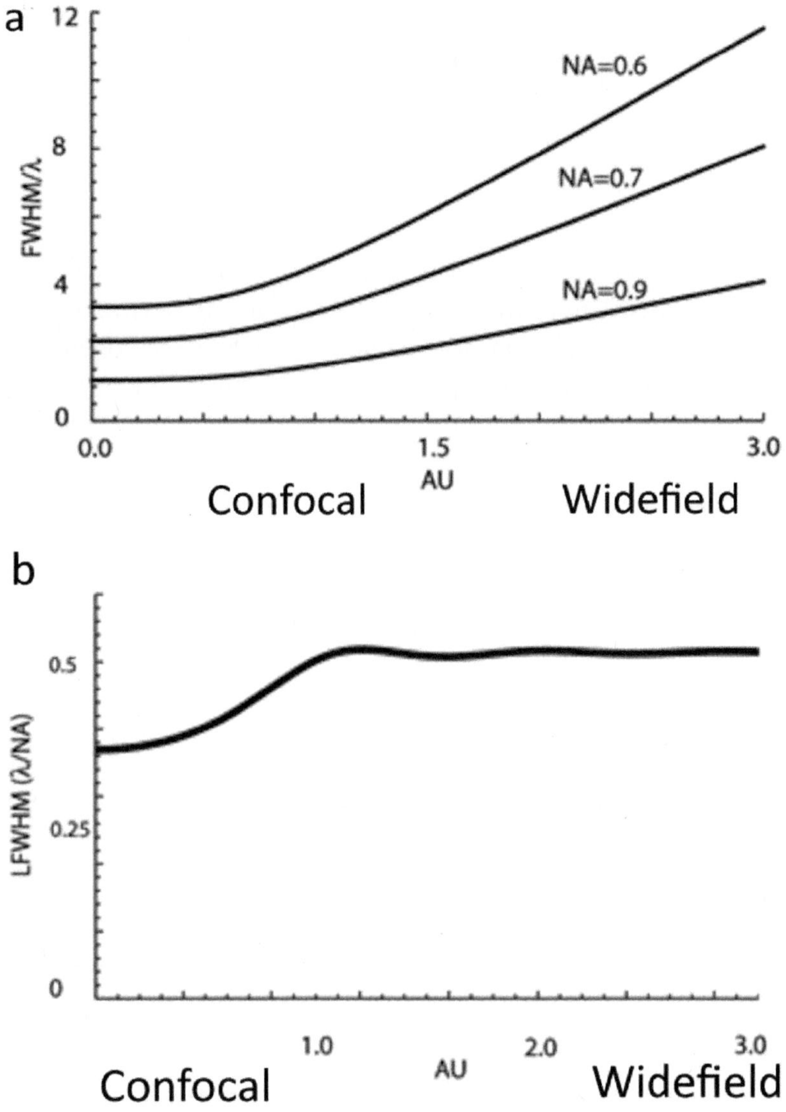

FIGURE 2.1   Axial (left panel) and lateral (right panel) responses (measured as full width at half maximum, FWHM) to pinhole diameter (measured in Airy Units, AU). The optical section thickness (axial resolution) is depicted for three different numeric apertures, *NA*. Modified from [8] with permissions.

The axial resolution δ_z of an OCT system in media with refractive index n equals the round-trip coherence length of the source, which is defined by the parameters of the light source [10]

$$\delta_{Z,OCT} = \frac{2ln2}{\pi n} \frac{\lambda_0^2}{\Delta\lambda_{FWHM}} \tag{2.11}$$

Here, λ_0 is the central wavelength, and $\Delta\lambda_{FWHM}$ is the spectral bandwidth of the light source defined in terms of full width at half maximum, FWHM.

Thus, the axial resolution can be optimized using low-coherence light sources with wider spectral bandwidth. That is why light-emitting diodes (LED), tungsten halogen lamps (THL), and superluminescent diodes (SLD) are the primary light sources in OCT.

The depth of focus (which is typically denoted by b in OCT) is also defined differently in terms of the beam waist $(\omega(z) \leq \sqrt{2}\omega_0$, where ω_0 is the radius of the beam waist)

$$b = \frac{2\pi n}{\lambda_0}\omega_0^2 = \frac{1}{2\pi}\frac{n\lambda_0}{NA^2} \tag{2.12}$$

The depth of focus (Eq.2.12) defines the range where the signal can be collected. Thus, each A-scan in OCT collects the signal from the focal volume defined by its width δ_{XY} and the axial extension b. Outside the focal volume, the intensity coming back from the sample is reduced considerably.

In retinal OCT, with a focal length of the eye of f eye = 16.7 mm and the refractive index $n_{vitreous}$ = 1.336, the lateral resolution is typically about 10 µm, resulting in a depth of focus of approximately 700 µm [9].

The typical axial resolution of OCT systems is 5–20 µm. The spectral width of the therapeutic windows (with low absorption in tissues) limits the maximum achievable resolution. For example, to achieve 2 µm axial resolution for 850 nm or 1,050 nm light sources used in ophthalmic systems, the bandwidth needs to be around 175 nm, which would exceed the width of the spectral window in the NIR range.

2.2 PHYSIOLOGICAL CONSIDERATIONS

Tissue is organized on multiple levels. Thus, the highest possible resolution is not always required. Instead, the required resolution can be tailored to a specific task.

We can consider three scales of interest: micro, meso, and macro. Firstly, it is at the cellular and subcellular levels. For example, changes to the organelles' sizes are typical for cancer transformations. In particular, it is known (see, for example, Appendix A) that the nuclei are enlarged in cancer cells.

Secondly, it is a meso or multicellular scale. For example, it can be an investigation of tissue morphology for cancer detection.

Finally, it is the macro scale, where the large body area (e.g., several to hundreds sq. cm) is imaged. It is the most common scenario for modalities other than microscopy. In this case, one particular area of interest will be to capture inhomogeneities in microcirculation.

Cellular and multicellular scales are well described in the literature. Here, we briefly consider the microcirculatory scale and its implications on imaging.

One can ask what the scale of changes seen in the body is. For example, what is the typical scale of changes in skin temperature?

These questions can be linked to several typical scales in microcirculation.

Firstly, it is an intercapillary distance. The oxygen diffusion, limited to 150 μm, determines this distance.

Thus, to image the capillary network, we need to sample the tissue with spatial resolution, which is less than half (Shannon–Nyquist criteria) of the typical intercapillary distance (150 μm): <75 μm.

However, it is a rough estimation. To make a more accurate estimation, we can consider the average surface density of capillaries for a particular body part. For example, skin has 16–65 capillaries per squared mm [11].

However, capillaries are not entirely independent units. They are supplied by the arterioles that occur at random intervals of 1.5–7 mm [12]. Each arteriole branches into four to five smaller branches that form the upper horizontal plexus of blood vessels. The blood vessel pattern in this network can be considered an umbrella shape, where the paired ascending arteriole and descending venule represent the handle. This "umbrella," comprising arterial and venous microvessels, is often called a "vascular unit." The vascular unit can be considered the smallest unit of microcirculation. In particular, one can expect that thermoregulation occurs at the vascular unit level.

Thus, if the imaging system needs to image large body areas and resolve temperature variations, then the imaging system may need to be able to resolve vascular units.

We can briefly consider the vascular units' implications on physiological optical imaging.

2.2.1 Tissue Model

We can build a cutaneous tissue model based on the vascular unit description. In particular, we can introduce the notion of "vascular zones" that cover the surface of the tissue. Each of these zones corresponds to a single vascular unit. Thus, the vascular zone is a 2D (surface) representation of a 3D object (vascular unit).

Each vascular zone can be perfused independently. Blood supply can be turned on/off using a putative sphincter (non-glabrous skin) or AV anastomoses (glabrous skin).

Vascular units do not need to be in direct contact with each other. They can be surrounded by a buffer zone supplied with oxygen and nutrients through diffusion from the vascular zone. We can denote the width of this area as Δ. This width is determined by the maximum length, which can be serviced by diffusion (around 150 μm). It should be noted that all skin cells are supplied through diffusion. However, the mentioned buffer zone (diffusion zone) does not have capillaries underneath. It can have a slightly different skin color, observed in some skin conditions (e.g., cutis marmorata or livedo reticularis).

We can represent this geometry using a lattice model, where the surface of the tissue is covered with tiles. Each tile consists of a vascular zone and a respective diffusion zone (see Figure 2.3). These tiles have random shapes and sizes. We will not make any assumptions about the tile shape other than that they mostly have convex shapes. In this case, if we denote the distance between centers of tiles as L, it can also serve as an estimation for the diameter of the tile (the diameter for the convex shape is the largest distance that can be formed between two opposite parallel lines tangent to its boundary). If the width of the diffusion zone for each vascular zone is equal to Δ, then the distance between the two vascular zones is 2Δ.

This lattice is imaged by a camera, consisting of a rectangular grid of pixels. Let's denote the pixel size at the tissue surface level as l. This geometry is represented in Figure 2.3. It should be noted that the pixel can be individual cell size or several cells binned together (binning is the grouping of outputs collected from several cells (pixels). It is an efficient way to increase sensitivity and SNR).

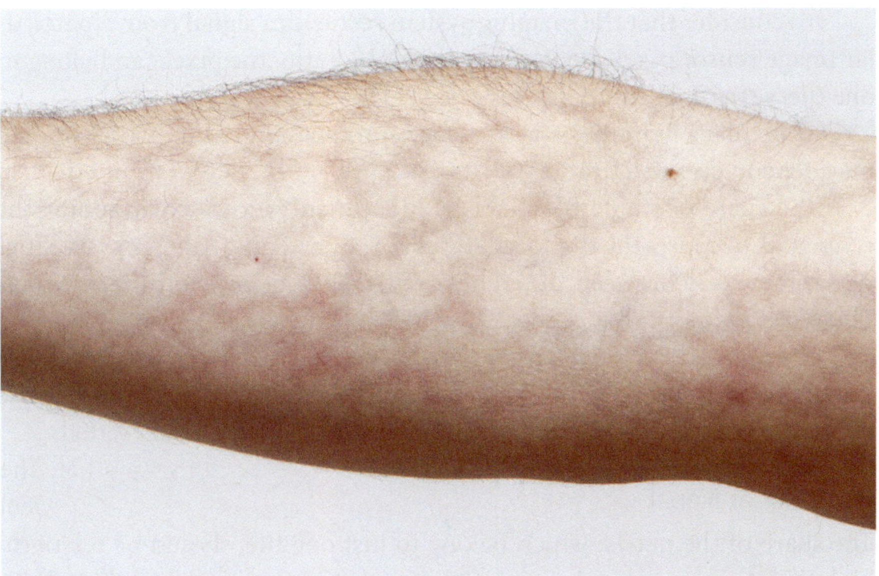

FIGURE 2.2   Mottled skin is often triggered by a decrease in temperature. Source: Stock.com/Liubomirt.

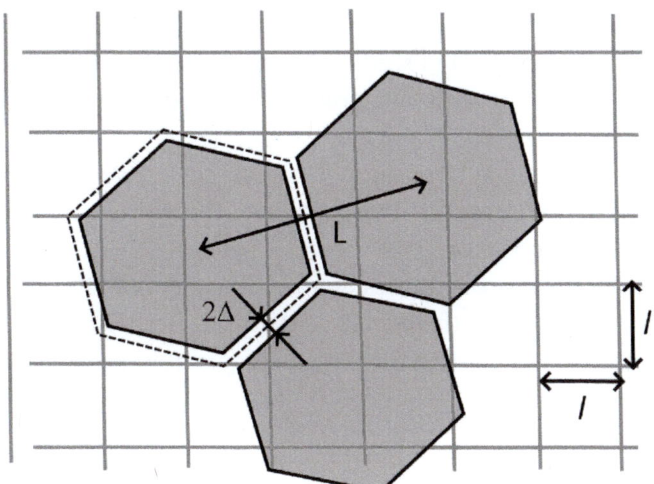

FIGURE 2.3   The cutaneous tissue imaging geometry. Grey hexagons depict vascular zones with intercenter distance L. The rectangular grid with size l illustrates camera pixels. Vascular zones are separated by a distance of 2Δ (diffusion zones). Oxygen supply in these areas is provided by diffusion from vascular zones. The dashed line depicts the diffusion zone for one vascular zone.

Let's consider that the imaging system retrieves a signal from a particular image sensor pixel. Depending on the l/L ratio, the pixel can belong to one tile, several, or multiple tiles.

Our primary hypothesis is that the reproducibility and quality of data may depend on the l/L ratio.

If the share of single-tile pixels is high, then we can expect each pixel's signal to represent the local status for a particular tile. However, the signal may vary between different pixels if they belong to different tiles. Therefore, we can call it a "local" signal.

If the pixel size is much larger than the tile size, then the pixel averages signals from multiple tiles. Thus, we expect this signal to represent the large skin patches in this scenario. So, we can call it a "global" signal.

2.2.2 Local Signal

The share of the pixels, which belong to just one tile, should be relatively high to get a proper local signal. Otherwise, the tentatively combined signals from several vascular zones may complicate interpretation.

We can consider several possibilities. Some pixels will be entirely within one vascular zone. It is an ideal situation, and we can expect the strongest signal. We will denote the share of such pixels as p_l.

If the pixel is small enough and covers (fully or partially) the diffusion zone, then the collected signal will be weaker (partial cover) or even absent (covers only the diffusion zone). We will denote the share of such pixels as p_0.

Finally, if the pixel is large enough, it can cover two or more vascular zones, which may complicate interpretation. Therefore, we will denote the share of such pixels as p_n.

To calculate these probabilities, we can consider one tile.

We can expect that if the center of the pixel lies within a certain distance x from the edge of the tile, it can image another vascular zone (p_n subset of pixels) if

$$X + \Delta \leq \frac{l}{\sqrt{2}} \tag{2.13}$$

Here, we used $l/\sqrt{2}$ (half diagonal) as the longest distance from the center of the pixel to any other pixel point. We have two scenarios here to consider. If the pixel is small enough, no pixels can image more than one vascular zone. For larger pixels, the inequality can be satisfied for some x.

Thus, we can write that the pixel images several vascular zones if its center lies within [0, X] from the tile edge, where

$$X = \begin{cases} 0 & \text{if } \quad l/\sqrt{2} \leq \Delta \\ l/\sqrt{2} - \Delta & \text{if } \quad \dfrac{l}{\sqrt{2}} > \Delta \end{cases} \tag{2.14}$$

Similarly, if the center of the pixel lies within a certain distance y from the edge of the tile, then it may image (fully or partially) the diffusion zone (p_0 subset of pixels) if the center of the pixel is located on the distance y from the edge so that

$$Y \leq \Delta + l/\sqrt{2} \tag{2.15}$$

Thus, we can write that the pixel images the diffusion zone (fully or partially) if its center lies within [0, Y] from the tile edge, where

$$Y = \Delta + l/\sqrt{2} \tag{2.16}$$

These three scenarios are depicted in the following table.

Probability	Description	Pixel center location (Distance measured from the edge)
p_0	Pixels that fully or partially sample diffusion zone	$\max\left(0, \dfrac{l}{\sqrt{2}} - \Delta\right) < X < \dfrac{l}{\sqrt{2}} + \Delta$
p_n	Pixels that sample several vascular zones	$X \leq \max\left(0, \dfrac{l}{\sqrt{2}} - \Delta\right)$
p_1	Pixels that fully sample just one vascular zone	$X > \Delta + l/\sqrt{2}$

We can notice that these scenarios are exhaustive. Consequently, $p_0 + p_1 + p_n = 1$.

These zones are schematically depicted in Figure 2.4.

Thus, the share of the potentially ambiguous pixels p_n or pixels with potentially weak signal p_0 can be estimated as the ratio of the area of the respective band (dS) to the whole tile (S) area. The band area can be calculated roughly as $dS = Pw$, where P is the perimeter of the tile and w is the band's width. However, it is accurate only for $l, \Delta << L$. In a more general

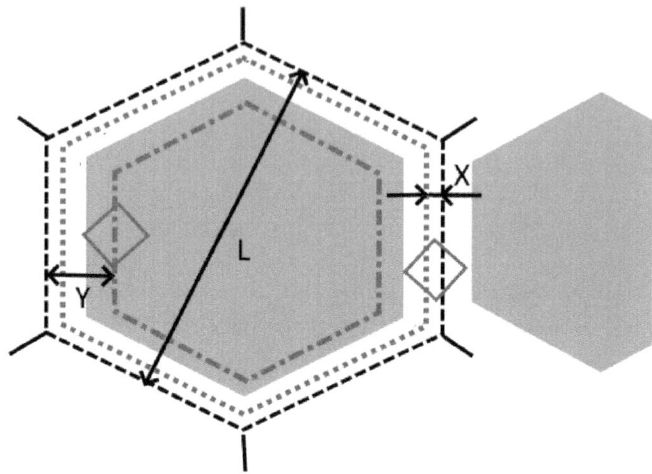

FIGURE 2.4   Grey hexagons depict the vascular zones. The dashed line illustrates the outer boundaries of the diffusion zone. The dash-dotted line represents the outer boundary of the area where the p_1 subset of pixels resides. The dotted line illustrates the boundary for the p_n subset of pixels.

case, the band area can be estimated using the assumption that the inner and outer boundaries are similar shapes. In this case, if the tile diameter L corresponds to tile area S, then the diameter of the inner boundary $L - 2l/\sqrt{2} - 2\Delta = L - \sqrt{2}l - 2\Delta$ corresponds to the inner area S'. Namely, $S/L^2 = S'/\left(L - \sqrt{2}l - 2\Delta\right)^2$. Thus, we can write

$$p_1 = \frac{S'}{S} = \begin{cases} \left(1 - \left(\sqrt{2}l + 2\Delta\right)/L\right)^2 & if \quad \sqrt{2}l + 2\Delta < L \\ 0 & if \quad \sqrt{2}l + 2\Delta > L \end{cases} \tag{2.17}$$

Similarly, we can get estimations for p_0 and p_n.

As mentioned, we will get the proper local signal if the share of p_1 is close to 1. However, it is not always possible. For example, in the mentioned case $\Delta = 150\,\mu m$, $L = 1.5\,mm$, p_1 cannot exceed 0.64. Thus, a significant share of pixels (>36%) potentially can have a convoluted signal from several vascular zones, which may complicate interpretation.

The share p_1 for several realistic scenarios is plotted in Figure 2.5.

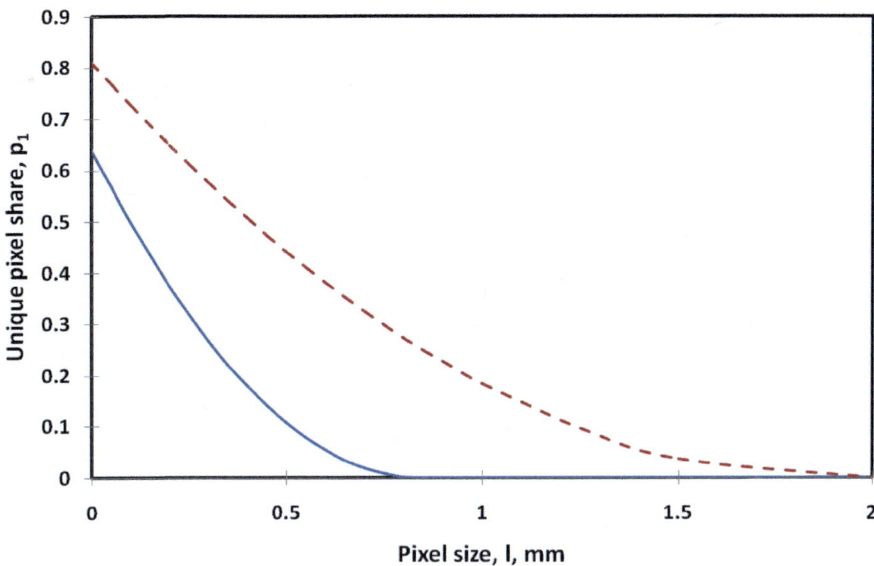

FIGURE 2.5   The unique pixel share, p_1, as a function of the pixel size l., Δ was set to 150 μm. L was 1.5 mm (solid blue line) and 3 mm (dashed red line).

2.2.3 Global Signal

The global signal will be representative of a large skin patch if it averages the signal from many vascular zones. This requirement can be written in terms of the ratio of pixel area to the tile area

$$\frac{t^2}{S} \gg 1 \tag{2.18}$$

The tile area can be estimated as $S = CL^2$, where C is a lattice-specific constant. We can deduce some insights about the constant C's value from specific geometries. For example, for the square lattice $C=1$. Thus, one can expect that C is close to 1 in the general case. Consequently, Eq.2.18 can be rewritten as

$$l^2 \gg L^2 \tag{2.19}$$

For practical purposes, Eq.2.19 can be reduced to the following rule of thumb: $l > 3L$ for the global signal collection.

2.3 SAFETY CONSIDERATIONS

For *in vivo* imaging, the system needs to comply with multiple safety standards. In particular, it needs to be safe from a photobiological point of view.

Light may pose multiple health hazards. As such, the medical device with light sources may represent the incidental hazard not just to the target area but to other body parts of a patient and health professional, for example, exposed skin and eyes. In particular, according to ISO 60601–2–57 Standard, light in the 200–300 nm range may pose the following hazards: actinic UV (180–400 nm), near UV (315–400 nm), blue light (300–700 nm), retinal thermal (380–1400 nm), corneal/lens IR (780–3000 nm).

The photobiological safety of medical devices is governed by several safety standards. If the device contains laser sources, it needs to meet requirements and be tested according to IEC 60601–2–22, Medical electrical equipment – Part 2–22: Particular requirements for basic safety and essential performance of surgical, cosmetic, therapeutic, and diagnostic laser equipment.

Suppose the device contains non-laser light sources (e.g., LEDs). In that case, it must be tested according to IEC 60601–2–57, Medical electrical equipment – Part 2–57: Particular requirements for the basic safety and essential performance of non-laser light source equipment intended for therapeutic, diagnostic, monitoring, and cosmetic/aesthetic use.

The photobiological safety of light sources is covered by IEC 62471: Photobiological safety of lamps and lamp systems.

How is it all relevant to resolution? In Chapter 3, we will see that the number of collected photons determines the signal-to-noise ratio (SNR). Thus, to achieve a certain SNR, each pixel (and corresponding area on tissue) needs to receive a certain number of photons. As we decrease this area (improving resolution), we need to increase the power density (irradiance) of illumination [W/m^2]. At some point, we may hit the safety limits imposed by safety standards, as emission limits are defined in power density (irradiance, [W/m^2]) or power density per solid angle (radiance, [W/m^2/sr]) terms. These limits are defined for each type of hazard.

2.4 WAYS TO IMPROVE RESOLUTION

As we saw, the imaging resolution is primarily limited by diffraction limit and pixel size. There are several approaches to tackle these limitations.

2.4.1 Immersion Optics

As discussed in section 2.1.1.1, the resolution of the optical system is determined by the numeric aperture, *NA*, of the lens (see Eq. 2.1).

In light microscopy, oil immersion is a technique used to increase a microscope's resolving power by increasing the objective lens's numerical aperture (NA) (see Eq.2.2).

For conventional microscopy (air ($n = 1$) as an immersion medium), the state-of-the-art *NA* is 0.95. The immersion microscopy covers the coverslip/specimen and objective lens with immersion fluid (oil). In the case of immersion oil, *NA* can be as high as 1.6.

Immersion oils are transparent oils with specific optical and viscosity characteristics necessary for use in microscopy. The oil's refractive index (RI) is typically chosen to match the RI of the microscope lens glass and the coverslip. Typical oils used have an RI of around 1.515. Synthetic oils are primarily used as they have several advantages over natural oils (cedar tree oil). They also do not exhibit autofluorescence.

In addition to increasing *NA*, immersion optics removes specular reflection on objective/air and specimen/air interfaces. It results in a larger amount of collected light and better contrast (in the case of reflection geometry).

2.4.2 Near-Field Imaging

The Rayleigh criterion (Eq.2.4) is applicable to the so-called far field, which refers to the distances far from the antenna. The electromagnetic fields on the short distances from the antenna are called near field. However, the boundary between the near field and far field regions is only vaguely defined, and it depends on the dominant wavelength (λ) emitted by the source and the size of the radiating element.

The distinction between them can be drawn by the radiation field strength decreases with distance. The far zone is characterized by inverse dependence of the field and inverse-square for power intensity. The near zone is characterized by more rapid decay. In particular, it is split into two subzones: reactive (closer to the antenna) and radiative (further from the antenna). The radiative field decreases with the squared inverse distance. In contrast, the reactive field decreases by an inverse-cube law, resulting in diminished power in the parts of the electric field by an inverse fourth power and sixth power, respectively.

In particular, for antennas equal to, or shorter than, half of the wavelength of the radiation, far field refers to the distances more than two wavelengths from the antenna. In contrast, the distances, which are less than one wavelength from the antenna, are referred to as near field.

Near-field scanning optical microscopy (NSOM) or scanning near-field optical microscopy (SNOM) uses evanescent or non-propagating near fields that exist in close proximity to the surface of the object. Because of this, the detector must be placed very close to the sample in the near-field zone, typically a few nanometers. Thus, a feedback mechanism is typically employed to maintain the probe at a certain distance from the surface.

In SNOM, the excitation laser light is focused through an aperture with a diameter smaller than the excitation wavelength, resulting in an evanescent field (or near field) outside the aperture. When the sample is scanned at a small distance below the aperture, the optical resolution of transmitted or reflected light is limited only by the diameter of the aperture. In particular, a lateral resolution of 6 nm [13] and a vertical resolution of 2–5 nm have been demonstrated [14]. However, due to the necessity of placing the probe very close to the surface, the applications of SNOM are mainly limited to surface analysis.

2.4.3 Sub-Pixel Resolution Techniques

As the pixel size is one of the primary limiting factors of optical imaging systems, several sub-pixel resolution approaches have been proposed. Current techniques are mainly based on acquiring several images with some changes to the scene (e.g., shifts) or illumination between them and following numerical reconstruction.

In particular, subpixel perspective sweeping microscopy [15] uses images taken at several illumination angles. Between two illumination angles, the shadows of the sample move over a subpixel distance. Then, a highly resolved image is reconstructed numerically.

Using the holographic approach, the resolution increase has been investigated using several subpixel-shifted holograms. The subpixel shifts are induced by changing the source's wavelength by several tens of nanometers [16] or the position of the source on top of the sample [17].

Rostykus et al. [18] demonstrated subpixel resolution for a compact lensless microscope. They created subpixel shifts by changing the injection current of a vertical-cavity surface-emitting laser (VCSEL). With pixel size 5.2 μm, they achieved a resolution of 2.76 μm over a FOV of ~28 mm^2:

Recently, Chen et al. [19] reported expansion microscopy (ExM). They demonstrated ExM with effective ~70 nm lateral resolution in cultured cells and brain tissue, performing three-color super-resolution imaging of ~10^7 $μm^3$ of the mouse hippocampus with a conventional confocal microscope. However, the technology requires synthesizing a swellable polymer network within a specimen.

NOTE

1 Note that multiple sub-pixel resolution techniques have emerged. They will be considered in Section 2.4.3.

REFERENCES

1. G Hollows, N James. "The airy disk and diffraction limit". www.edmund optics.ca/knowledge-center/application-notes/imaging/limitations-on-resolution-and-contrast-the-airy-disk
2. www.infosmore.com/blog/how-to-read-smartphone-camera-specs-the-ultimate-guide/
3. CE Shannon. "Communication in the presence of noise". *Proc Inst Radio Eng* 37(1): 10–21 (1949).
4. G Saiko, A Pandya, I Schelkanova, et al. "Optical detection of a capillary grid spatial pattern in epithelium by spatially resolved diffuse reflectance probe: Monte Carlo verification". *IEEE J Sel Top Quantum Electron* 20(2): 187–195 (2014).
5. www.vision-doctor.com/en/camera-technology-basics/sensor-and-pixel-sizes.html
6. www.med.unc.edu/microscopy/wp-content/uploads/sites/742/2018/06/clsm-tutorial-v2.pdf
7. www.microscopyu.com/microscopy-basics/depth-of-field-and-depth-of-focus
8. T Wilson. "Resolution and optical sectioning in the confocal microscope". *J Microscopy* 244: 113–121 (2011).
9. S Aumann, S Donner, J Fischer, F Müller. "Chapter 3: Optical coherence tomography (OCT): Principle and technical realization". In *High Resolution Imaging in Microscopy and Ophthalmology: New Frontiers in Biomedical Optics [Internet]*, JF Bille, Ed. Cham: Springer (2019). www.ncbi.nlm.nih.gov/books/NBK554044/
10. AF Fercher, W Drexler, CK Hitzenberger and T Lasser. "Optical coherence tomography – principles and applications". *Rep Prog Phys* 66(2): 239–303 (2003).
11. NC Wetzel, Y Zotterman. "On differences in the vascular colouration of various regions of the normal human skin". *Heart* 13: 357 (1927).

12. IM Braverman, J Schechner. "Contour mapping of the cutaneous micro-vasculature by computerized laser Doppler velocimetry". *J Invest Derrnatol* 97(6): 1013–1028 (1991).
13. X Ma, Q Liu, N Yu, et al. "6 nm super-resolution optical transmission and scattering spectroscopic imaging of carbon nanotubes using a nanometer-scale white light source". *Nat Comm* 12(1): 6868 (2021).
14. Y Oshikane, T Kataoka, M Okuda, et al. "Observation of nanostructure by scanning near-field optical microscope with small sphere probe (free access)". *Sci Techn Adv Mat* 8(3): 181 (2007).
15. G Zheng, SA Lee, Y Antebi, et al. "The ePetri dish, an on-chip cell imaging platform based on subpixel perspective sweeping microscopy (SPSM)". *Proc Natl Acad Sci USA* 108: 16889 (2011).
16. W Luo, Y Zhang, A Feizi, et al. "Pixel super-resolution using wavelength scanning". *Light Sci Appl* 5: e16060 (2016).
17. W Luo, A Greenbaum, Y Zhang, et al. "Synthetic aperture-based on-chip microscopy". *Light Sci Appl* 4: e261 (2015).
18. M Rostykus, M Rossi, C Moser. "Compact lensless subpixel resolution large field of view microscope". *Opt Lett* 43(8): 1654–1657 (2018).
19. F Chen, PW Tillberg, ES Boyden. "Optical imaging. Expansion microscopy". *Science* 347(6221): 543–548 (2015).

CHAPTER 3

Contrast Ratio in Physiological Optical Imaging

IN THIS CHAPTER, WE will consider the primary aspects of image quality. We start with technical (section 3.1) and physiological (section 3.2) considerations. Then, we consider common ways to improve image quality (section 3.3).

3.1 TECHNICAL CONSIDERATIONS

The primary purpose of virtually any physiological imaging technique is to extract and quantify a physiologically relevant parameter: heart rate, tissue blood saturation, etc.

Thus, as the first step, we need to establish a reference point for the analytical ability of a method. Then, we need to link it to the camera properties. Then, we will establish a common approach to evaluate image quality using a contrast ratio. We also consider common factors that reduce image quality (specular reflections).

3.1.1 Limits of Detect and Quantification

In engineering, the quality of signal acquisition is measured in terms of signal-to-noise ratio (SNR).

However, what SNR is sufficient enough for diagnostic purposes? We can get insights into it from analytical methods. In analytical methods

DOI: 10.1201/9781003482505-4

(including clinical laboratory tests), there are two relevant levels of signal quantification: limit of detection (LOD) and limit of quantification (LOQ) (see, for example, [1]).

The limit of detection is defined as an analyte content that can be distinguished from the blank (no analyte) with an error probability of (1-β). Here β is the likelihood of false-negative results: type II error.

The limit of quantification is defined as an analyte content that can be determined with a certain (often arbitrary) level of precision.

LOD and LOQ can be defined using the mean and variance of the signal, which can be combined in a single number metric, signal-to-noise ratio (SNR). Typically, LOD and LOQ are defined using SNR with values of 3 (or 3.3) and 10 for LOD and LOQ, respectively (see, for example, [2]).

Figure 3.1 illustrates the relationship between the blank, the limit of detection (LOD), and the limit of quantification (LOQ) by showing the probability density function for normally distributed measurements at the blank, at the LOD defined as 3x standard deviation of the blank, and at the LOQ defined as 10x standard deviation of the blank. For a signal at the LOD, the alpha error (probability of false positive) is small (1%). However,

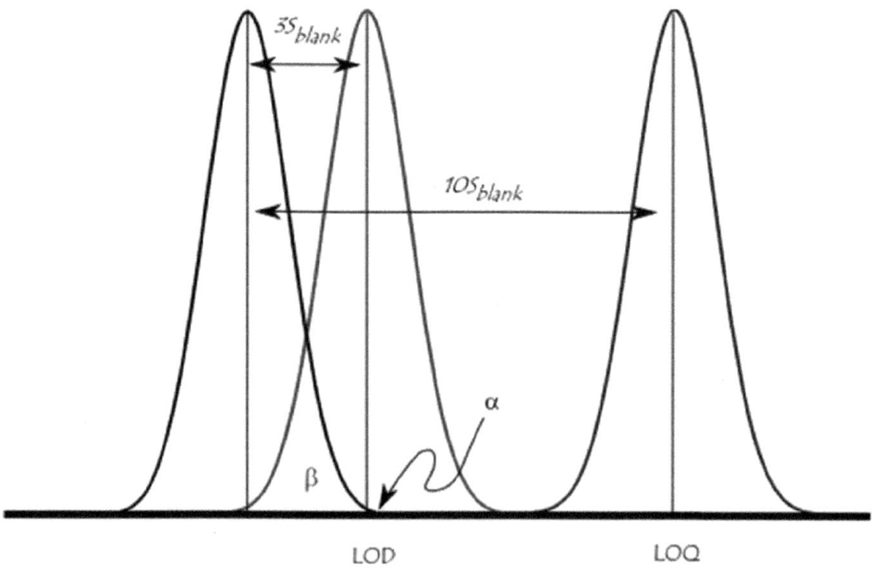

FIGURE 3.1 Illustration of the concept of detection limit and quantification limit by showing distributions associated with blank, detection limit (LOD), and quantification limit (LOQ) level samples. (Reproduced from [3] under CC BY-SA 4.0 license).

the β error (probability of a false negative) is 50% for a sample with a concentration at the LOD. This means a sample could contain an impurity at the LOD, but there is a 50% chance that a measurement would give a result less than the LOD. At the LOQ, there is minimal chance of a false negative.

We are interested in quantitative measurements, so that we will use LOQ with SNR = 10 for further discussion.

3.1.2 CMOS Camera Noise

To assess SNR, we need to identify the mean and variability of the signal. The variability of the signal is caused by noises. In particular, the CMOS camera has several types of noise: shot noise, dark noise, and quantization noise.

The shot noise can be modeled by a Poisson process. In this case, the variance of the noise σ_p^2 is equal to the mean number of photons μ_p. The dark noise consists of read noise and dark current. However, the dark noise can be ignored in most practical cases other than low light conditions (e.g., astronomy). The quantization noise arises from digitizing the continuous voltage signal into a digital one and can be modeled by a Gaussian distribution. The read noise, dark current, and quantization noise are set values for a particular camera and typically can be found in spec sheets.

We can take into account that sources of noise are independent and use error propagation rules to write expressions for the camera output mean and SNR:

$$\mu_y = K\left(\mu_d + \eta\mu_p\right) \tag{3.1}$$

$$SNR = \frac{\eta\mu_p}{\sqrt{\sigma_d^2 + \sigma_q^2 / K^2 + \eta\mu_p}} \tag{3.2}$$

Here, K is the camera's sensitivity; η is the sensor's quantum efficiency (wavelength dependent), μ and σ refer to mean and standard deviations for photons (subindex p), dark noise (subindex d), quantization (subindex q), and output (subindex y).

However, for simplicity, we will ignore dark noise and quantization noise. This assumption is generally valid for all cases other than low-lit conditions. In this case, the SNR can be written as (here subscript 1 refers to the single-pixel SNR):

$$SNR_1 = \frac{\eta\mu_p}{\sqrt{\eta\mu_p}} = \sqrt{\eta\mu_p} \tag{3.3}$$

3.1.3 Contrast Ratio

To facilitate rapid technology development, there is a need for standard metrics/approaches that can be used for fast technology comparison and assessment. The contrast ratio can potentially become a basis for such an approach. While being introduced more than 190 years ago by Weber and Fechner [4], the contrast ratio is mainly overlooked in the current tissue imaging literature. Instead, various alternatives (e.g., dynamic contrast ratio) have been proposed. However, Weber's simple definition has numerous advantages. For example, Saiko et al. have shown its utility for biomedical imaging applications [5].

The contrast ratio provides a convenient way to analyze images. The contrast ratio at any point <x,y> on the surface of the tissue can be defined according to Weber's law as

$$c(x, y) = \frac{\varphi_b - \varphi(x, y)}{\varphi_b} \tag{3.4}$$

where φ_b and $\varphi(x, y)$ are signals at some distant point (background) and any point of interest $<x, y>$, respectively.

Although the minimal contrast ratio for the human eye is around 0.1 (for brightness response, it is approximately 0.14 for cones and 0.015–0.03 for rods [6]), images with a lower contrast ratio can be digitally enhanced and still be used for feature examination or pattern recognition via image processing.

The related question is, what contrast ratio can be detected by a camera?

If we consider illumination by a narrow-band light source, the signal will most likely be detected by a single channel (whenever it is an RGB or monochrome camera). Thus, the relevant parameter is the bit depth of the camera, d. Typically, it is 8 bits for consumer-grade cameras and 12 for scientific-grade cameras. The minimal contrast, which can be captured by a camera with bit depth d, can be estimated as [5]

$$c_{min} = \frac{1}{2^d - 1} \tag{3.5}$$

This expression is derived for a linear sensor. However, a similar equation can be derived for a power law sensor (e.g., gamma correction case).

Thus, the linear sensor's minimal contrast ratio is close to 0.4% for consumer-grade cameras ($d = 8$) and 0.025% for scientific-grade cameras ($d = 12$).

3.1.4 Specular Reflection

Specular reflection is one of the primary types of light-tissue interaction. It arises from a mismatch of the refractive indexes on the interface between two media. Specular reflection is weakly dependent on the wavelength of incident light and can, therefore, occur similarly across the spectrum. For the normal incidence, the coefficient of specular reflection R_s will depend on the relative index of refraction n:

$$R_s = \frac{(n-1)^2}{(n+1)^2} \tag{3.6}$$

Typical values of R_s for biological tissues/air interface are relatively low, in the range of a few percent, because typical tissue will exhibit refractive index n in the range of 1.33–1.45 [7]. Despite this low value of specular reflection, such measurement is problematic as a noise source for the other light-tissue interactions due to its propensity to create hot spots.

Specular reflection depends not only on the refractive index but also on the roughness of the surface. Typical surfaces are not perfectly smooth and, therefore, exhibit some roughness through surface components meeting at varying angles (per Figure 3.2a), where the components' size dictates the surface's roughness from the smooth surface to closer to Lambertian surface [8]. The variation in angles causes angular distribution in specular reflection, as is common in matte surface finishing of metal or plastic. Surfaces that may visually appear to be smooth may contain roughness that isn't visually perceptible, such as skin.

In virtually any application, the specular reflection is an undesirable artifact that masks the useful signal. Interestingly, specular reflection

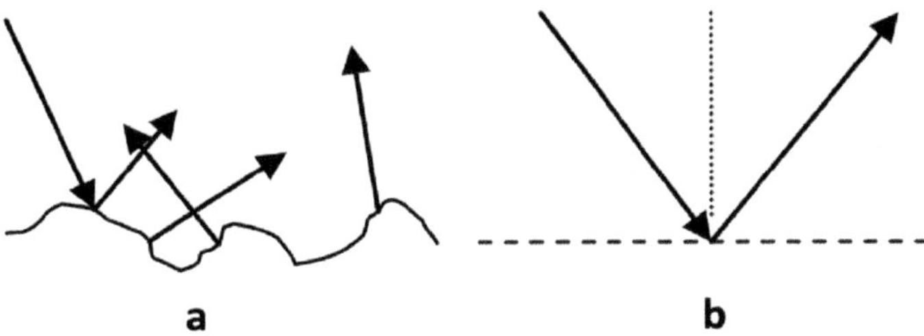

FIGURE 3.2 Two types of specular (Fresnel) reflection from tissue. Specular reflection from (a) a rough surface (i.e., Lambertian reflectance) and (b) a smooth surface.

is disruptive while relatively weak compared to diffuse reflectance. For example, diffuse reflection measured in reflection mode PPG and rPPG is typically around 20–60% (depending on the wavelength), much stronger than only several percent for specular reflection [9]. The outsized effect of specular reflection compared to diffuse reflection is explained by the concentration of specular reflection within a narrow solid angle, whereas diffuse reflection is distributed across a large (2π) angle. Specifically, specular reflection can overwhelm diffuse reflection with concentrated specular reflection at certain angles. When occurring in imaging applications (such as rPPG), this phenomenon is known as "hot spots" [10] on images, where specular reflection locally saturates the image sensor.

However, the "hot spots" are not the only issues introduced by the specular reflection in physiological optical imaging. As most physiological optical imaging modalities aim to collect information from deeper tissues (e.g., diffuse reflectance), the specularly reflected light deteriorates the contrast by mixing with the light from deeper tissues and increasing the denominator in Eq.3.4 (the nominator includes the difference of two light fluxes and will not be impacted by specular reflection addition).

3.2 PHYSIOLOGICAL CONSIDERATIONS

The realistic physiological signal can be pretty faint. For example, the amplitude of the remote PPG signal (see Figure 3.3) can be on a scale of less than 1% of the average value (background). Thus, in this case, the useful signal is the modulation, not the background itself. To illustrate it, let's suppose that the mean of the background signal is μ. If we consider its slight change in time $\mu' = \mu - \Delta$, then the SNR for physiological modulation can be found as

$$SNR_{phys} = \frac{\eta\mu - \eta\mu'}{\sqrt{\eta\mu}} = \frac{\eta\Delta}{\sqrt{\eta\mu}} = \frac{\Delta}{\mu} SNR_{background} \qquad (3.7)$$

Consequently, to quantify the physiological signal, we need to meet LOQ (namely, SNR at least 10:1) for the signal on the scale Δ/μ of the background ("blank" in analytical methods terminology). Thus, for our example using Eq.3.7, we can estimate it as $SNR = 10 = 0.01\sqrt{\eta\mu_p}$, which requires around 10^6 photons per measurement.

This scenario is unattainable for many sensors. For example, the pixel's full-well capacity can be on the scale of 10,000-20,000 photons. The ways to overcome it will be discussed in section 3.3.1.

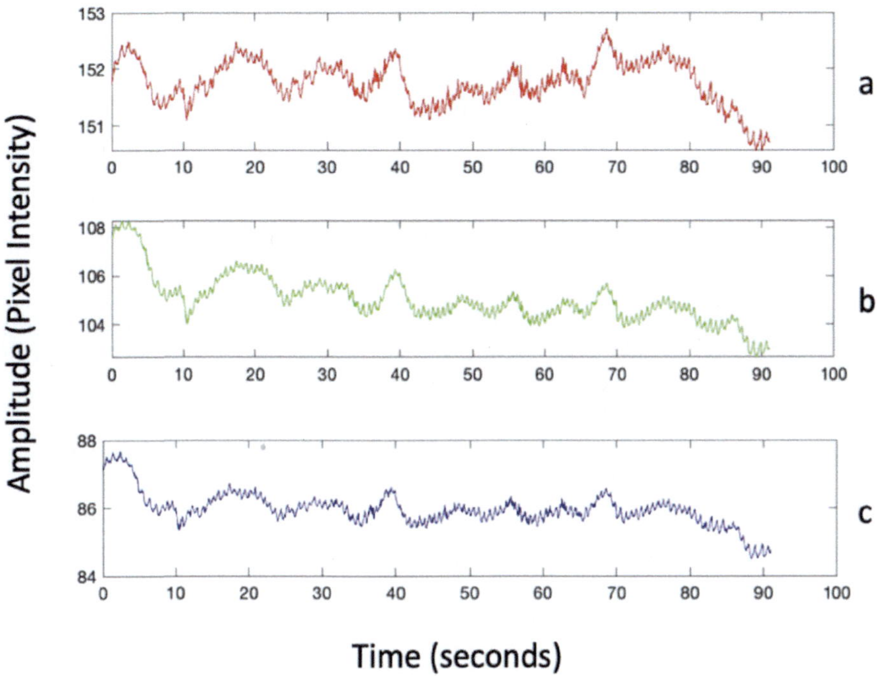

FIGURE 3.3   An example of a remote PPG signal. Each subplot represents raw data collected by red (a), green (b), and blue channels (c) of the RGB camera, respectively. Low-frequency (0.1 Hz) oscillations are visible in addition to the cardiac pulsations.

Alternatively, instead of temporal variation (PPG signal), we can consider spatial changes in reflectance (e.g., capillary visualization). In this case, Δ/μ in Eq.3.7 can be substituted with the contrast ratio.

3.3 WAYS TO INCREASE IMAGING CONTRAST

As improving the image contrast is a recurring issue in many optical modalities, multiple contrast enhancement techniques have been developed.

3.3.1 Binning

As we calculated in section 3.2, physiological imaging may require 10^6 photons per point measurement, which is unattainable for many sensors. For example, the pixel's full-well capacity can be on the scale of 10,000-20,000 photons. Fortunately, there is a way to achieve this by using binning.

Binning refers to aggregating signals from several nearby pixels (spatial binning) or the same pixel over several consecutive frames (temporal binning or image stacking).

The concept of binning lies in the fact that the measured signal adds linearly with the number of observations while the noise grows proportionally to the square root of the number of measurements (assuming its independence):

$$SNR_N = \frac{\sum_N \eta \mu_p}{\sqrt{\sum_N \eta \mu_p}} = \frac{N \eta \mu_p}{\sqrt{N \eta \mu_p}} = \sqrt{N}\sqrt{\eta \mu_p} = \sqrt{N} \times SNR_1 \qquad (3.8)$$

Thus, binning over N measurements leads to improvement in SNR on a scale of $\sqrt{N}$.

3.3.1.1 Spatial Binning

Spatial binning is grouping outputs collected from several cells (pixels). It is an efficient way to increase sensitivity and SNR. In CCD devices, binning can be achieved on the sensor level. For CMOS devices, the binning cannot be achieved on the sensor level; it can be done digitally. It is less efficient than sensor-level binning (for example, it cannot reduce read noise); however, it provides significant SNR improvement anyway.

In most scenarios, binning is achieved by averaging signals over *NxN* cells. Taking into account that sources of noise in each cell are independent, we can use error propagation rules and write similarly to Eq.3.8

$$SNR_{N \times N} = N \times SNR_1 \qquad (3.9)$$

Thus, *NxN* binning allows approximately *Nx* improvement in SNR in the CMOS camera.

For example, for 10x10 binning, the required number of photons per cell in the example from section 3.2 will be on the scale of 10^4, which is more plausible. However, the specific value for binning size N must be determined from the relative magnitude of the expected signal and camera parameters.

Also, it should be noted that *NxN* binning reduces the image's resolution by the factor of *Nx*.

3.3.1.2 Temporal Binning or Image Stacking

Temporal binning or image stacking is a widespread technique in astro-photography. It takes multiple images of the same stellar object, often during several nights. Then, these images are processed to obtain one with a better signal-to-noise ratio and enhanced dynamic range. In this case, according to Eq. 3.8., binning over N measurements leads to improvement in SNR on a scale of $\sqrt{N}$. The critical step in image stacking is image registration.

3.3.1.3 Ensemble Averaging

Both spatial and temporal binning have their disadvantages. In particular, spatial binning leads to resolution degradation. Temporal binning is not suitable for imaging dynamic objects.

However, it is possible to develop a special temporal binning technique for periodic processes called "ensemble averaging." The idea is to take a long signal, cut it into segments one period long, align them, and average over several cycles. So, to some extent, it is similar to image stacking for the series (or ensemble) of frames. Obviously, this concept can be extended to per pixel in the imaging modalities.

This approach works not only for periodic signals but also for quasi-periodic signals like heartbeats and pulse propagation. The use of ensemble averaging for the analysis of remote photoplethysmographic (rPPG) signals has been proposed by Kamshilin et al. [11]. They cut the PPG waveform into individual PPG pulses so that the beginning of each pulse coincided with the corresponding R-peak of the ECG, and then these pulses were ensemble averaged. The concept is illustrated in Figure 3.4. In the original work, Kamshilin et al. [11] used a mean waveform obtained by averaging the filtered one-cardiac-cycle oscillations during 30 s to estimate pulse wave velocity (PWV) in the carotid basin. In the follow-up work [12], the same group averaged PPG pulse over 12 subsequent cardiac cycles to monitor changes in tissue blood perfusion during abdominal surgery.

3.3.2 Narrow Band Imaging

Narrow-band imaging is a popular method to increase imaging contrast of absorption heterogeneities, such as blood vessels. It is widely used in many medical fields. In particular, it is quite popular in endoscopy (see Chapter 11 for further details).

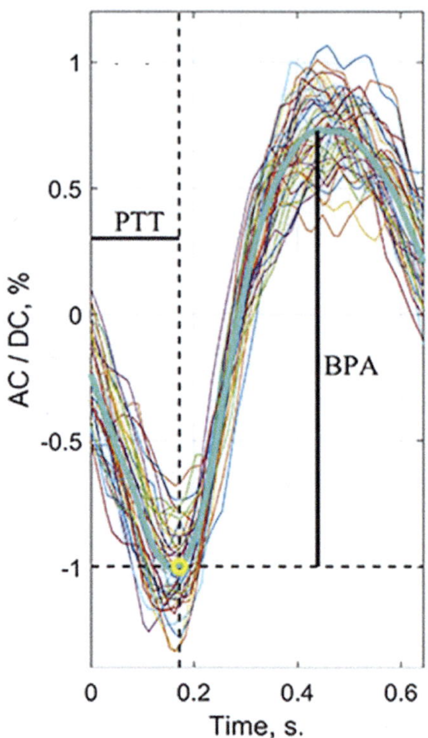

FIGURE 3.4   One-cardiac-cycle waveform (thick green line) was obtained after averaging particular waveforms (thin color lines) over 30 cycles. Reproduced from [11] under CC BY 4.0 license.

The concept of narrow-band imaging is to illuminate the tissue with light with a relatively narrow spectral range, typically centered around the absorption band of the target chromophore. As the most commonly targeted chromophore in biomedical imaging is hemoglobin, typically narrow-band imaging includes illumination in blue (Soret band) or green (Q band) spectrum ranges.

We can estimate what kind of improvement can be brought by narrow-band imaging. In particular, for illustration purposes, we can consider a homogeneous media (characterized by absorption coefficient μ_α and scattering coefficient μ_s) with a defect (e.g., a blood vessel) characterized by different absorption or scattering coefficients. The tissue is imaged in the reflectance mode.

We can start with the standard definition of the contrast ratio $c = (I - I_d)/I$, where I and I_d are the light intensities of the background and feature (defect), respectively. Considering that reflected light is the product of external illumination (which can be assumed constant within a small area) and the reflectance of the tissue, we can move from the intensity of reflected light (I) to the reflectance of the tissue (R). If we consider the tissue heterogeneity (e.g., capillary) as absorption (or scattering) perturbation $\Delta\mu$, then in the first-order approximation, we can write

$$c = \frac{I - I_d}{I} = \frac{R - R_d}{R} = \frac{-\Delta R}{R} = \frac{-1}{R}\frac{\partial R}{\partial \mu}\Delta\mu \qquad (3.10)$$

Here μ denotes a coefficient of absorption (for an absorption defect) or a coefficient of scattering (for a scattering defect).

To proceed further, we need to specify the model of light propagation in the tissue. To illustrate the approach, we can use the two-flux Kubelka–Munk model (see, e.g.[13]), a well-known approach in tissue optics that provides reasonable estimations in a wide range of conditions. For the semi-infinite slab, the solution for the tissue reflectance R is well known (see Appendix B):

$$R = x + 1 - \sqrt{(x+1)^2 - 1} \qquad (3.11)$$

Here $x = k/s$, where k and s are the tissue absorption and scattering per unit length, respectively. For diffuse illumination, $k = 2\mu_a$ and $s = 2\mu_s$ (where μ_a and μ_s are tissue absorption and scattering coefficients). Taking into account that $\dfrac{\partial R}{\partial \mu_a} = \dfrac{\partial R}{\partial x}\dfrac{\partial x}{\partial \mu_a}$ and $\dfrac{\partial x}{\partial \mu_a} = \dfrac{1}{\mu_s}$ one can write $c_a = \dfrac{(-1)}{R}\dfrac{\partial R}{\partial \mu_\alpha}\Delta\mu_\alpha = \dfrac{x}{\sqrt{(x+1)^2 - 1}}\dfrac{\Delta\mu_\alpha}{\mu_\alpha}$ for the contrast ratio of an absorption defect. Expressing x as a function of R from Eq.3.11: $x = (1-R)^2 / 2R$ and substituting it into the last equation gives us

$$c_a = \frac{(1-R)}{(1+R)}\frac{\Delta\mu_a}{\mu_a} \qquad (3.12)$$

Similarly, for the scattering defect:

$$c_s = -\frac{\left(1-R\right)}{\left(1+R\right)}\frac{\Delta\mu_s}{\mu_s} \tag{3.13}$$

Eqs.3.12 and 3.13 are explicit relationships between the relative absorption (scattering) coefficient changes and the contrast ratio. Tissue reflectance R can be measured experimentally.

A trivial consequence of Eqs.3.12 and 3.13 is that absorption and scattering defects have the opposite impact on the contrast ratio. Increased absorption leads to reduced intensity (positive contrast). Increased scattering leads to increased intensity (negative contrast).

Narrow-band techniques at absorption maximums may have better (2x–3x) sensitivity than broadband ones due to the contribution of two factors: increased relative change in the absorption coefficient and increased amplification factor (due to increased absorption). Wavelength selection is very important: the increase in absorption leads to a decrease in reflectance and an increase in sensitivity.

In Fig.3.5, one can see the results of emulation of the spectral dependence of visualization of an absorption defect (background: 3% fully

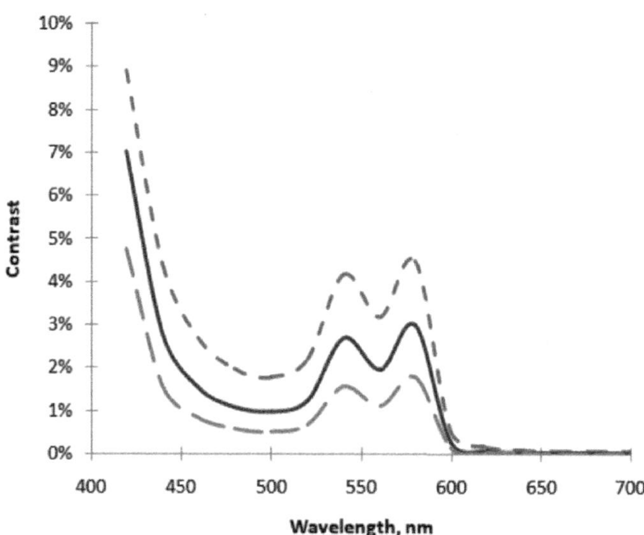

FIGURE 3.5 Spectral dependence of the contrast ratio of an absorption defect for various skin types (light (dash line), medium (solid line), and dark (long dash line)).

oxygenated blood; defect: 3.3% fully oxygenated blood, which corresponds to a 10% increase in the blood content) for various skin types [14]: light skin tone (4% melanin content), medium skin tone (12% melanin content) and dark skin (30% melanin content). The epidermis thickness was assumed to be one-third of the probing depth, approximately 0.5 mm in the visible spectrum [8]. One can see that the contrast is maximal at absorption peaks.

3.3.3 Suppression of Specular Reflection

As mentioned earlier, specular reflection often represents a significant challenge in imaging by decreasing the contrast and adding "hot spots." Despite the issues that specular reflection may cause, they are not insurmountable. Over the course of years, several approaches to mitigating it have been developed.

3.3.3.1 Optically Shielding

Design options to overcome specular reflection include optically shielding the light source from the detector so that skin specular reflections with high angular density do not enter the detector and overwhelm the relatively lower-strength scatter signal. However, it is not a universal solution. In particular, rPPG is more vulnerable to specular reflection than the traditional spatially resolved mode PPG due to the increased difficulty of integrating optical shielding.

3.3.3.2 Polarization Gating

In addition to the control measures already described, polarization gating is a practical approach to identify and isolate specular reflection [15, 16]. Among multiple approaches to isolate the image-bearing photons (time gating, spatial filtering, and coherence gating, [17]), polarization gating has the highest appeal because of the relative simplicity of instrumentation, making it a potential tool amenable for clinical use.

The polarization gating is based on the observations that when polarized light transverses a tissue, it gradually depolarizes due to the multiple scattering/refraction events in the inhomogeneous (scattering) medium (i.e., tissue). However, the polarization state of specularly reflected or weakly scattered light returned from the tissue surface retains its original polarization state. Thus, cross-polarization optical schema should effectively eliminate the specular reflection contribution (polarization filtration or polarization gating).

Karlsson et al. [18] investigated polarization filtration's utility in mitigating motion artifacts in Laser Doppler Perfusion Imaging. They found that polarization filtration reduces the perfusion signal ($p < 0.001$). They also found that movement artifacts were lower ($p < 0.05$).

Sidorov et al. [19] investigated the image enhancement of photoplethysmographic imaging using polarization optics. They compared the blood pulsation amplitude measured with and without polarization filtration and found that the mean difference between the pulsation amplitude measured under polarized and non-polarized illumination is 3.1 ± 1.7%. Therefore, they concluded that the influence of polarization filtration on the readout of information from cutaneous vessels pulsating at the heartbeat frequency is insignificant. However, they also found that polarization filtration efficiently washes out the high-frequency noise related to movement artifacts and improves the shape of the PPG waveform, allowing the identification of the parameters of cardiac cycles with higher reliability.

Kamshilin et al. [20] used polarization filtration in combination with video photoplethysmography to investigate the utility of blood pulsation amplitude in characterizing the relative change of the cutaneous blood.

The specular reflection issue arises again in polarization imaging with two orthogonal channels. To address this issue, Jacques et al.[21] proposed an optical schema where the glare from the air/glass/skin interfaces was removed through optical coupling (using index matching liquid) of a glass plate in contact with the skin and by off-normal illumination. Morgan and Stockford [22] proposed a method that combines images obtained with linearly and circularly polarized light to produce a polarization-gated image free from surface reflections and does not require optically flat plates or matching fluid.

3.3.3.3 Time Gating

With recent advances in imaging sensors, time gating has also become an affordable technique. Various manufacturers developed time-of-flight (ToF) imaging sensors. As a result, ToF systems started penetrating biomedical research. For example, Zhao et al. [23] investigated the ToF utility for diffuse optical tomography (DOT) applications. They reduced acquisition time by implementing confocal measurements and multiplexing the illumination sources. In particular, ToF systems can remove glare caused by surface reflection by rejecting ultrafast photons.

3.3.4 Other Methods

As image contrast enhancement is a recurring theme across different optical modalities, multiple techniques have been developed. Here, we briefly consider several such approaches.

3.3.4.1 Lock-in

Lock-in detection extracts a signal from a noisy background by multiplying its input with a reference signal. A lock-in amplifier employs a homodyne detection scheme and low-pass filtering to measure a signal's amplitude and phase relative to a periodic reference. The technology has achieved quite remarkable results. For example, state-of-the-art lock-in amplifiers can measure a signal in the presence of noise up to a million times higher in amplitude than the signal of interest (120 dB dynamic reserve) [24].

The following observation can illustrate the lock-in detection principle of operations.

$$<\sin(\omega t)\sin(\omega_r t)> = \frac{2}{T}\int_0^T \sin(\omega t)\sin(\omega_r t)\,dt = \begin{cases} 1, & \omega = \omega_r \\ 0, & \omega \neq \omega_r \end{cases} \quad (3.14)$$

Here, $<...>$ denotes averaging over time interval T, ω is the signal angular frequency, and ω_r is the angular frequency of the reference source.

Lock-in detection has been known for almost a century [25, 26]. However, for many years, this quite bulky and expensive instrument was confined to scientific laboratories as a single or several-channel piece of scientific equipment. With advances in imaging sensors and computing systems, the technology has recently been extended to imaging modalities. In this case, the lock-in detection happens on a per-pixel basis. Since then, several optical lock-in modalities emerged. For example, lock-in thermal imaging is widely used in the nondestructive testing of materials.

Some lock-in optical modalities have emerged for biomedical applications. Most of them use external periodic excitation of the test object. For example, Bonmarin and Gal [27] developed a lock-in thermal imaging setup for dermatological applications. Their apparatus uses a temperature-modulated airflow to stimulate the skin surface periodically. They detected small variations of the tissue thermophysical properties, including perfusion variations, for benign skin lesions. The same group also performed numerical simulations using a two-dimensional axisymmetric multilayer heat-transfer model to investigate the feasibility of using lock-in thermography for early-stage detection of cutaneous melanoma [28].

However, it is possible to extend the lock-in detection to passive modalities (without external signals of a known frequency) or quasi-periodic signals, such as a cardiovascular pulse. For example, Kamshilin et al. [29] developed an optical lock modality for photoplethysmographic imaging. They derived a reference function for synchronously detecting cardiovascular pulse waves from the same frames.

3.3.4.2 Digital Transformations

So far, we have considered hardware-based methods for image quality improvements (although lock-in detection can have a fully digital implementation, for example, [29]). However, numerical methods can also be used. In particular, image contrast enhancement (ICE) is vital in various image processing and computer vision applications. As such, it is a very active research area.

Contrast enhancement methods can be divided into spatial and transform domain enhancements [30]. Spatial domain contrast enhancements are based on grayscale transformations of nonlinear functions, such as logarithmic transformation [31], gamma function [32], histogram-based technology [33], and nonlinear secondary filtering [34]. Transform domain contrast enhancements are based on filtering methods [35]. Histogram equalization is the most used spatial domain image enhancement method. An adaptive histogram equalization (AHE) method was proposed in [36]. This method estimates the probability density of the original input image. Then, the scale factor is adjusted adaptively according to the average intensity value of the image. A popular version of the adaptive histogram equalization approach is Contrast Limited AHE (CLAHE) [37]. CLAHE is a variant of adaptive histogram equalization in which contrast amplification is limited to reduce this noise amplification problem. In CLAHE, the slope of the transformation function gives the contrast amplification in the vicinity of a given pixel value.

REFERENCES

1. Clinical and Laboratory Standards Institute. *Protocols for Determination of Limits of Detection and Limits of Quantitation, Approved Guideline.* CLSI (2004).
2. S Lister. "Validation of HPLC methods in pharmaceutical analysis". In *Separation Science and Technology [Internet].* Elsevier, pp. 191–217 (2005).

https://linkinghub.elsevier.com/retrieve/pii/S0149639505800510 (cited February 14, 2022)

3. https://en.wikipedia.org/wiki/Detection_limit

4. GT Fechner. *Elemente der Psychophysik. band 2.* Leipzig: Breitkopf und Härtel (1860).

5. G Saiko, A Douplik. "Contrast ratio quantification during visualization of microvasculature". *Adv Exp Med Biol* 1072: 369–373 (2018).

6. www.telescope-optics.net/eye_intensity_response.htm (accessed July 5, 2024)

7. Y Mendelson, BD Ochs. "Noninvasive pulse oximetry utilizing skin reflectance photoplethysmography". *IEEE Trans Biomed Eng* 35(10): 798–805 (1988).

8. R Lu. *Light Scattering Technology for Food Property, Quality and Safety Assessment.* CRC Press (2017).

9. A Douplik, G Saiko, I Schelkanova, V Tuchin. "The response of tissue to laser light". In *Lasers for Medical Applications. Diagnostics, Therapy and Surgery*, H Jelinkova, Ed. Woodhead Publishing, pp. 47–109 (2013).

10. K Tan, X Cheng. "Specular reflection effects elimination in terrestrial laser scanning intensity data using Phong model". *Remote Sens* 9(8) (2017).

11. AA Kamshilin, TV Krasnikova, MA Volynsky, et al. "Alterations of blood pulsations parameters in carotid basin due to body position change". *Sci Rep* 8: 13663 (2018).

12. AA Kamshilin, VV Zaytsev, AV Lodygin, VA Kashchenko. "Imaging photoplethysmography as an easy-to-use tool for monitoring changes in tissue blood perfusion during abdominal surgery". *Sci Rep* 12(1): 1143 (2022).

13. G Saiko, A Douplik. "Reflectance of biological turbid tissues under wide area illumination: Single backward scattering approach". *Int J Photoenergy* 241364 (2014).

14. SL Jacques. "Origins of tissue optical properties in the UVA, visible and NIR regions". In *Advances in Optical Imaging and Photon Migration*, Vol. 2, RR Alfano, JG Fujimoto, Eds. Washington, DC: OSATOPS: Optical Society of America, pp. 364–371 (1996).

15. RT Tan, K Nishino, K Ikeuchi. "Separating reflection components based on chromaticity and noise analysis". *IEEE Trans Pattern Anal Mach Intell* 26(10): 1373–1379 (2004).

16. MJ Leahy, JG Enfield, NT Clancy, et al. "Biophotonic methods in microcirculation imaging". *Med Laser Appl* 22(2): 105–126 (2007).

17. JC Hebden, SR Arridge, DT Delpy. "Optical imaging in medicine: 1. Experimental techniques". *Phys Med Biol* 42: 825–840 (1997).

18. MGD Karlsson, K Wårdell. "Polarized laser Doppler perfusion imaging--reduction of movement-induced artifacts". *J Biomed Opt* 10(6): 064002 (2005).

19. IS Sidorov, MA Volynsky, AA Kamshilin. "Influence of polarization filtration on the information readout from pulsating blood vessels". *Biomed Opt Express* 7: 2469–2474 (2016).

20. AA Kamshilin, AV Belaventseva, RV Romashko, et al. "Local thermal impact on microcirculation assessed by imaging photoplethysmography". *Biol Med (Aligarh)* 8: 361 (2016).

21. SL Jacques, K Lee. "Polarized video imaging of skin". *Proc SPIE* 3245: 356–362 (1998).
22. SP Morgan, IM Stockford. "Surface reflection elimination in polarization imaging of superficial tissue". *Opt Lett* 28(2): 114–116 (2003).
23. Y Zhao, A Raghuram, HK Kim, et al. "High resolution, deep imaging using confocal time-of-flight diffuse optical tomography". *IEEE Trans Pattern Anal Mach Intell* 43(7): 2206–2219 (2021).
24. Zurich Instruments MFLI. www.zhinst.com/products/mfli-lock-in-amplifier (accessed August 2, 2024).
25. CR Cosens. "A balance-detector for alternating-current bridges". *Proc Phys Soc* 46: 818 (1934).
26. WC Michels, NL Curtis. "A pentode lock-in amplifier of high frequency selectivity". *Rev Sci Instrum* 12: 444 (1941).
27. M Bonmarin, FA Le Gal. "A lock-in thermal imaging setup for dermatological applications". *Skin Res Technol* 21(3): 284–290 (2015).
28. M Bonmarin, FA Le Gal. "Lock-in thermal imaging for the early-stage detection of cutaneous melanoma: A feasibility study". *Comput Biol Med* 47: 36–43 (2014).
29. AA Kamshilin, S Miridonov, V Teplov, et al. "Photoplethysmographic imaging of high spatial resolution". *Biomed Opt Express* 2(4): 996–1006 (2011).
30. X Zhang, Y Ren, G Zhen, et al. "A color image contrast enhancement method based on improved PSO". *PLoS One* 18(2): e0274054 (2023).
31. B Olena, V Roman, I Ivasenko. "Color image enhancement by logarithmic transformation in fuzzy domain". In *2019 IEEE 2nd Ukraine Conference on Electrical and Computer Engineering*, pp. 1147–1151 (2019).
32. H Shih-Chia, C Fan-Chieh, C Yi-Sheng. "Efficient contrast enhancement using adaptive gamma correction with weighting distribution". *IEEE Trans Image Process* 22: 1032–1041 (2013).
33. W Yu, C Qian, Z Baomin. "Image enhancement based on equal area dualistic sub-image histogram equalization method". *IEEE Trans Consum Electron* 45: 68–75 (1999).
34. C Gao, K Panetta. "A new color contrast enhancement algorithm for robotic applications". *IEEE Int Conf Technol Pract Robot Appl*: 42–47 (2012).
35. M Junwon, J Yuneseok, N Yoojun, K Jaeseok. "Edge-enhancing bi-histogram equalisation using guided image filter". *J Vis Commun Image Represent* 58: 688–700 (2019).
36. PM Narendra, RC Fitch. "Real-time adaptive contrast enhancement". *IEEE*: 655–661 (1981).
37. SM Pizer, EP Amburn, JD Austin, et al. "Adaptive histogram equalization and its variations". *Comput Vision Graph Image Process* 39: 355–368 (1987).

CHAPTER 4

Sampling Depth in Physiological Optical Imaging

OPTICAL DIAGNOSTIC METHODOLOGIES RELY on light propagating through the tissue, collecting information about optical tissue properties, which can be further deciphered into physiological or diagnostic information. As tissues typically have a multilayered heterogeneous structure, it is important to understand which part of the tissue was sampled. For imaging modalities, the sampling volume can be considered a volume sampled by one pixel. Generally, the sampling volume can be characterized by the resolution (e.g., pixel size) and sampling depth. As we discussed the resolution and pixel size already in Chapter 2, we will turn our attention to the sampling depth in this chapter.

In this chapter, we will consider several models that help provide insights into the sampling depth. We also consider several common ways to increase the sampling depth.

4.1 TECHNICAL CONSIDERATIONS

To get insights into the sampling depth, we will start with a related concept: penetration depth.

DOI: 10.1201/9781003482505-5

4.1.1 Penetration Depth

The penetration depth measures how deep light or any other electromagnetic radiation can penetrate a media. The penetration depth is defined as the depth where light intensity falls to the level *1/e* of its initial power.

The penetration depth in the visible and near-infrared ranges of the spectrum is depicted in Figure 4.1.

The penetration depth depends not just on the wavelength but also on the type of illumination. If the tissue is illuminated with a collimated light source (e.g., laser beam or optical fiber), then the incident beam intensity in the tissue follows the Beer–Lambert law. Namely, it decays exponentially with distance z as $exp(-\mu_t z)$. Here, μ_t is the total attenuation coefficient ($\mu_t = \mu_a + \mu_s$). Thus, the penetration depth for collimated light will be $\delta = 1/\mu_t$. However, the light intensity of the diffuse light will follow the exponential law with $exp\left(-\mu_{eff} z\right)$. Here, $\mu_{eff} = \sqrt{3\mu_a\left(\mu_a + \mu_s'\right)}$. The same applies to diffuse illumination (e.g., ambient light). Thus, the penetration depth for diffuse light will be $\delta = 1/\mu_{eff}$.

4.1.2 Sampling Depth

Unlike the penetration depth, the sampling depth is a more vaguely defined term. Typically, it refers to the depth from which the signal is collected. As it depends on the collection geometry, we will briefly discuss it for typical geometries for optical diagnostics modalities: reflectance or transmittance geometries.

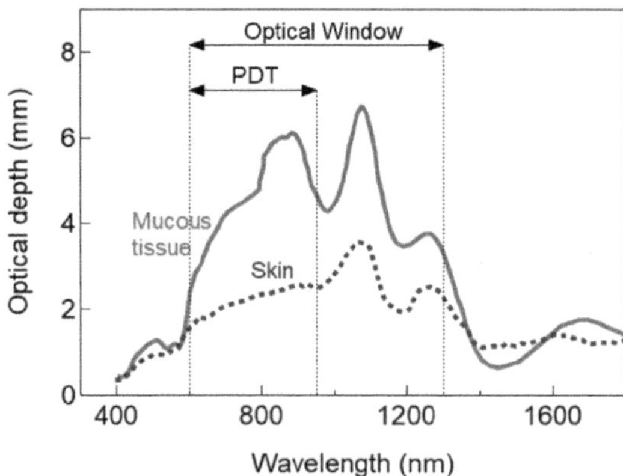

FIGURE 4.1   The penetration depth in the optical range of the spectrum for skin and mucosa. (Reproduced from [1] under CC BY 4.0 license).

4.1.2.1 Transmittance Geometry

In transmittance geometry, the source and detector are placed on different sides of the tissue. Given those restrictions, transmittance geometry applications are typically limited to protruding body parts, like the finger or earlobe. However, it is the most widely used clinical modality, as pulse oximetry is a transmittance-based modality. In addition, the transmittance mode can also be used for larger body parts, like the head, in functional near infrared spectroscopy (fNIRS).

For transmittance-based modalities, the light collects information from all tissue layers between the source and detector. However, the optical path can deviate from a straight line as light experiences scattering. For example, for a pulse oximeter, the light passes through the nail blade, nail bed, muscle, phalanx bone, adipose tissue, dermis, and epidermis of the plantar side of the hand. In this case, the light experiences significant scattering in the epidermis, dermis, and phalanx bone. In general, simple rules can be deduced as follows: more scattering leads to a longer mean optical path. Similarly, stronger absorption shortens the mean optical path, as ballistic and quasi-ballistic paths have higher contributions.

4.1.2.2 Reflectance Geometry

For reflectance-based geometry, the light source and detector are placed on the same side of a body part. As it does not have restrictions on the body part thickness, it can be applied virtually to any patch of skin or mucosa.

The light propagation can be characterized by the mean optical path (MOP) and the mean sampling depth (MD).

For reflectance-based geometry, there are two important cases to consider: spatially resolved spectroscopy and imaging geometries.

4.1.2.2.1 Spatially Resolved Spectroscopy   As the name implies, in the case of spatially resolved spectroscopy, the source and detector are separated in space (e.g., muscle oximetry). It is the most common case for contact modalities, like photoplesythmography (PPG). In this case, the light enters the tissue typically as a narrow beam (e.g., through an optical fiber). Similarly, the detector (or detectors) collects light from small skin patches located at predetermined distances from the sources. In this case, the light path between the source and detector has a characteristic "banana" shape.

Chatterjee et al. [2] found that in reflectance mode pulse oximetry (660 and 940 nm wavelengths), the mean optical path (MOP) and the mean sampling depth (MD) depend almost linearly on the source-detector separation. In particular, MOP and MD for red and IR channels are identical until 6 mm and 5 mm, respectively. After that, red and IR paths start diverging. Thus, for space-separated geometries, MOP and MD are determined primarily by geometrical factors, namely the source-detector distance, d. In particular, the sampling depth for this configuration is proportional to $\sqrt{2}d/4$ in the weak absorption limit ($d << \delta$) and $\sqrt{d\delta/2}$ in the strong absorption limit ($d >> \delta$) [3]. Here δ is the light penetration depth. Thus, these probes are designed to probe a particular tissue depth/layer (or layers).

4.1.2.2.2 Imaging Geometry   In a typical imaging scenario, there is no source-detector separation per se. In typical imaging applications, the tissue is illuminated by a broad beam (or ambient light), and the data is collected from a relatively large skin area. In this case, the light propagation can also be characterized by the mean optical path (MOP) and mean sampling depth (MD). However, unlike spatially resolved geometry, they are determined solely by the optical properties of the tissue. For example, in this case, the sampling depth is close to the light penetration depth δ. MOP can be estimated as 7.8δ [4]. As such, the sampling depth may vary for different wavelengths. Similarly to the transmittance mode, simple rules can be deduced: more scattering leads to a longer mean optical path, and stronger absorption shortens the mean optical path.

4.1.3 Related Concepts

There are several concepts related to the sampling depth.

4.1.3.1 Optical Density or OD

Optical density is a standard measure of optical transparency for optical filters, but it is occasionally used to characterize tissue transparency. Optical density, OD, can be defined as $OD = \mu_t d$, where d is the thickness of the layer. However, optical density is typically defined as $log_{10}\left(I_{in}/I_{out}\right)$, where I_{in} and I_{out} are intensities of incoming and outcoming beams, respectively (note that these definitions differ only by the factor of $ln10$). Thus, OD4 means that the filter's transmission is 10^{-4}. Optical density is broadly used in microbiology to characterize bacterial concentrations in inoculums.

4.1.3.2 Maximal Depth of Visibility

In addition, for imaging geometries, we can define the maximum depth of visibility of a certain inhomogeneity (e.g., capillary in capillaroscopy or dermoscopy). This depth can be linked to a contrast ratio resolved by the optical system. Saiko et al. [5] showed that the contrast ratio of the defect buried on the depth z follows the exponential law with $exp\left(-2\mu_{eff}z\right)$ (see Appendix B).

4.1.4 Single Backward Scattering

We can illustrate the calculation of certain metrics like mean optical path and mean sampling depth using a Single Backward Scattering developed in [6]. For convenience, some results are presented here. The more complete version is presented in Appendix B.

Let's consider that each scattering event occurs on the distance $l_s = 1/\mu_s$. If forward and almost forward scattering events dominate substantially, we can consider the following quasi 1D model (the path is almost ballistic; however, it can slightly deviate in transverse directions): if the photon was scattered n times forwardly (each with probability $1 - p$) and then was scattered backwardly (with probability p), then to reach the surface again it must undergo additionally n forward scattering events. The probability of this n-1-n path is $p(1-p)^{2n}$. However, we have not yet considered that the photon flux also undergoes absorption during this path.

To take absorption into account, let us consider the photon flux propagation between two scattering events. The intensity of outcoming flux $I_{out} = I_{in}exp(-\mu_a l_s) = I_{in}exp(-\mu_a/\mu_s)$. So, we can attribute an absorption multiplier (propagator) $a = exp(-\mu_a/\mu_s)$ to each pathway between two nearby scattering events.

Thus, the total contribution of all paths with single backward scattering (the coefficient of reflectance in the single backward scattering model) will be provided by the infinite series:

$$R_1 = pa^2 + p(1-p)^2 a^4 .... + p(1-p)^{2n} a^{2n+2} + ... \tag{4.1}$$

It is easy to calculate this sum if we take into account that it is a geometric series. As $\left|(1-p)^2 a^2\right| < 1$ the series converges and its sum is equal to

$$R_1 = \frac{pa^2}{1-(1-p)^2 a^2} \tag{4.2}$$

Using this approach, it is easy to calculate several statistics.

4.1.4.1 Mean Optical Path

Using the developed model, we can easily calculate the mean optical path (MOP).

For this purpose, we can notice that we can associate the average length between two scattering events $l_s = 1/\mu_s$ with each propagator a. Thus, if we take into account that each term in Eq.4.1 represents a probability of a respective optical path, we can write

$$<l> = 2l_s pa^2 + 4l_s p(1-p)^2 a^4 \ldots + (2n+2)l_s p(1-p)^{2n} a^{2n+2} \ldots \quad (4.3)$$

This infinite sum can be found if we notice that $<l> = l_s a \dfrac{\partial R_1}{\partial a}$. Then, differentiating the Eq.4.2, we can get

$$<l> = \frac{2pa^2 l_s}{\left(1 - (1-p)^2 a^2\right)^2} \quad (4.4)$$

Which can be expressed through Eq.4.2 as

$$<l> = \frac{2l_s R_1^2}{pa^2} \quad (4.5)$$

As a is close to 1, and taking into account that in 1D model $p = (1-g)/2$, we can roughly estimate $<l> = 4l_s' R_1^2$. Here, $l_s' = 1/\mu_s'$. More accurate estimation can be obtained using expansion $\varepsilon = 1 - g$ in Eq.4.2. Then, p can be estimated as $p = \varepsilon(1/\sqrt{2} - 1/2) \approx 0.2(1-g)$. Thus, a more accurate assessment gives us $<l> \approx 10l_s' R_1^2$

4.1.4.2 Penetration Depth

By construction of the model, the inbound flux on the depth nl_s (just below nth scattering event) is equal to $I_{n,\,in} = (1-p)^n a^n$. In addition, we have an outbound flux, which will be equal to $I_{n,\,out} = R_1 I_{n,\,in}$.

Thus, the penetration depth in terms of scattering events can be written as $1/e = (1-p)^n a^n$.

As a result, the penetration depth can be found as

$$\delta = -\frac{l_s}{\ln((1-p)a)} = \frac{1}{\mu_a - \ln(1-p)\mu_s} \quad (4.6)$$

Using a 1D model, this expression can be estimated as $\delta = \dfrac{1}{\mu_a + \mu'_s/2}$.

A more accurate assessment using Eq.4.2 gives us $\delta \approx \dfrac{1}{\mu_a + 0.2\mu'_s}$.

4.1.4.3 Sampling Depth

By definition, the expression for tissue reflectance in a single backward scattering approximation (Eq 4.1) is the sum of an infinite number of terms. We can compare this expression with the contribution to the reflectance of the first m-terms (scattering layers):

$$R_{1,m} = \frac{pa^2}{1-(1-p)^2 a^2}\left(1-(1-p)^{2m}a^{2m}\right) \tag{4.7}$$

So, if we limit ourselves by some threshold criteria (say we are interested in $1 - \alpha = 95\%$ of total contribution), we can easily calculate the number of the most contributing m-terms:

$$(1-p)^{2m}a^{2m} = \alpha \tag{4.8}$$

Thus,

$$m = \frac{\ln(\alpha)}{2\ln(a(1-p))} = \frac{\ln(\alpha)}{2(-\mu_a/\mu_s + \ln(1-p))} \approx \frac{-\ln(\alpha)}{2(\mu_a/\mu_s + p)} \tag{4.9}$$

In Figure 4.2, we can see the dependence of light penetration depth as a function of μ_a/μ_s for different α and g.

Multiplying Eq.4.9 on the scattering length $l_s = 1/\mu_s$, the sampling depth (SD) for the parameter α can be estimated as $SD(\alpha) = -\delta \ln(\alpha)/2$. For example for $\alpha = 0.05$, $SD(0.05) \approx 1.5\delta$.

So, it is easy to see that in the imaging modality, information is being gathered from a relatively thin surface scattering layer, which is not deeper than 0.5–1 mm, while the major part of information propagates from the top 2–3 transport depths (0.1 mm). Therefore, 2–5 upper scattering layers (m-terms) contribute more than 50% of the total reflected flux. Also, the higher the absorption (or the lower g), the higher the contribution of the upper layers.

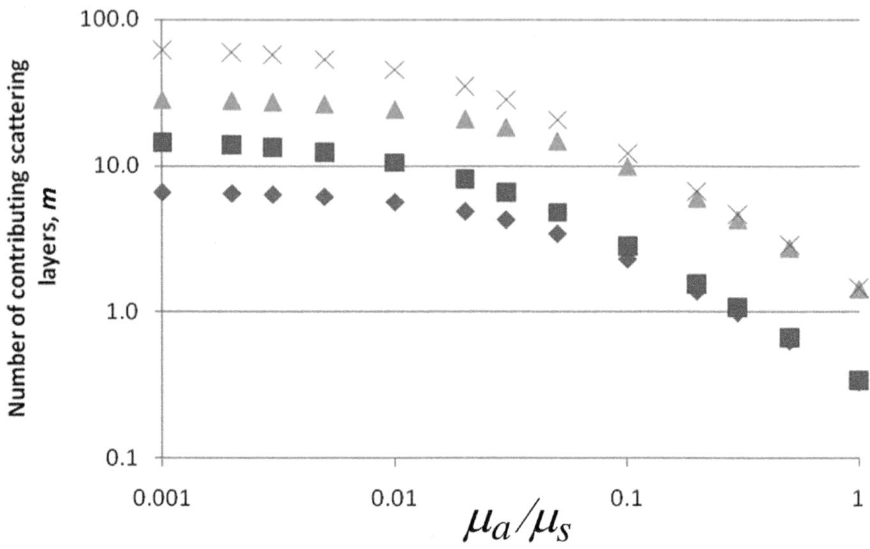

FIGURE 4.2 The number of contributing m-terms (scattering layers or transport depths) as a function of μ_a/μ_s ratio for different α and g. $\alpha = 0.5$, g $= 0.8$ (rhomboids), $\alpha = 0.5$, g $= 0.9$ (squares), $\alpha = 0.05$, g $= 0.8$ (triangles), $\alpha = 0.05$, g $= 0.9$ (crosses).

4.2 PHYSIOLOGICAL CONSIDERATIONS

As we mentioned, optical diagnostic modalities are uniquely positioned to investigate the properties of the body's superficial layers: skin and mucosa.

We can illustrate several common themes using the skin as an example. Similar considerations can be used for mucosa.

The skin consists of multiple layers, which, on a high level, can be grouped into the epidermis, dermis, and subcutaneous tissues. The epidermis and dermis are of primary interest for many clinical applications, particularly for wound care and oncology, which are the primary focus of this book.

From that perspective, the information on structural and functional properties of the epidermal and dermal layers is of great importance.

Different optical modalities focus on different aspects. For example, optical coherence tomography (OCT) can provide information on tissue morphology.[1] However, most imaging modalities provide functional information.

4.2.1 Epidermis

The epidermis is the avascular outermost protective layer of the skin. The epidermal layer is a relatively thin layer for most body parts. For hairy skin, it is 0.1 mm thick. For glabrous (non-hairy) skin, it is 1.5 mm thick.

As the epidermis is a relatively thin layer, its visualization is not a significant challenge, and many optical modalities can visualize it. Two primary absorbers here are melanin (visible range of spectrum) and water (IR and terahertz spectrum ranges).

Epidermal and epithelial layers are of primary interest for oncology as many carcinomas start there.

4.2.2 Dermis

The dermal layer is located under the epidermis layer. It is a 1 to 4 mm thick layer comprising connective tissue, blood and lymphatic vessels, hair follicles, nerves, glands, etc. However, mostly, it is about 2 mm thick.

This layer is significant for many clinical applications. One of the primary areas of interest is microcirculation, which consists of two primary plexuses: the superficial papillary plexus, which is located in the papillary dermis and serves the epidermis, and the reticular plexus located above the junction between the dermis and hypodermis to serve the hair follicles and sweat glands.

Microvasculature, and particular capillaries, are of great interest for various clinical applications, including wound care and oncology.

In particular, wound healing is associated with sufficient blood supply. Thus, direct (laser Doppler) and indirect (like tissue oxygen saturation) measures are used to characterize blood perfusion. Similarly, in oncology, tumor growth is typically associated with angiogenesis. In this case, the blood network becomes more dense and disorganized. These changes can be visualized using structural and functional imaging.

However, dermis imaging is a much more challenging task than imaging of epidermis. In the visible spectrum range, there is a significant absorption from blood components (hemoglobin). In addition, there is significant water absorption in the IR and terahertz range (see Figure 4.3). As the body consists of 65–70% water, it virtually limits the use of wavelengths longer than 2.3 μm for dermal imaging *in vivo* or *ex vivo* on unprocessed (e.g., thinly cut) samples.

Thus, the optical range for dermis investigations is mainly limited to visible and NIR spectrum ranges. As the shortwave NIR has deeper

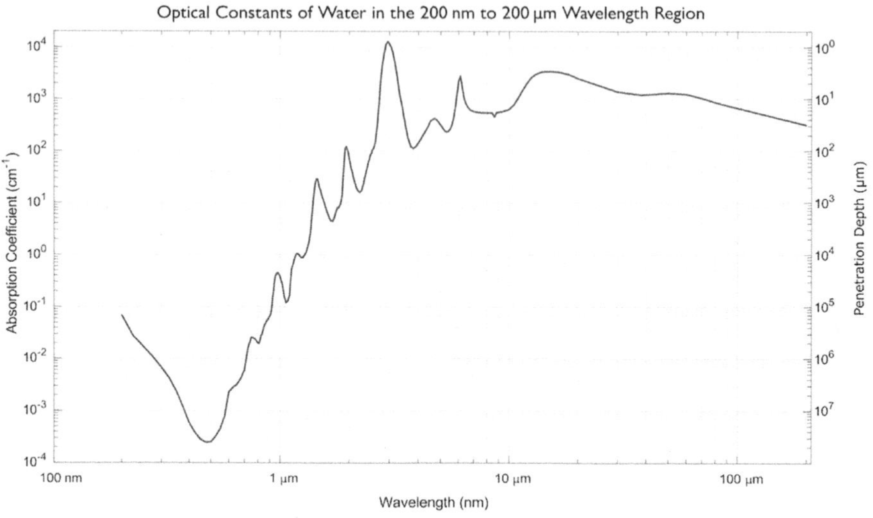

FIGURE 4.3   Absorption spectrum and penetration depth of water. Image courtesy of Faraz Sadrzadeh-Afsharazar. Plotted from data by Hale and Querry [7].

penetration depth, it is a preferred spectral range in structural imaging modalities, like optical coherence tomography (OCT).

In functional imaging, both visible and NIR ranges are used. As functional imaging typically collects the volume-averaged information instead of optical slicing, the use of visible and NIR ranges in functional imaging like hyperspectral/multispectral imaging (HSI/MSI) may have interesting implications as these ranges may probe different vascular networks. The shallow penetration depth of light in the visible spectrum range (see Figure 4.1) limits the sampling to the superficial papillary plexus. However, shortwave NIR has deeper penetration on a scale of several mm (see Figure 4.1). Thus, it samples both the superficial papillary plexus and the reticular plexus. These differences and their implications on HSI/MSI will be discussed in section 5.3.2.

Although water absorption is a severe limiting factor for multiple imaging modalities, some modalities (like terahertz imaging) exploit water absorption to detect water content alterations, which can be used for certain diagnostic purposes.

4.3 WAYS TO INCREASE SAMPLING DEPTH

Here, we consider several ways to increase the sampling depth. However, it should be noted that in many cases, improving contrast, which we considered in Chapter 3, and increasing the sampling depth are very interrelated problems. In particular, one can increase the contrast by increasing the sampling depth and vice versa. For example, increasing the contrast for an inhomogeneity located at a certain depth will most likely increase the sampling depth. Thus, the separation between contrast-enhancing and sampling depth-enhancing techniques is quite vague, and most of the techniques used for contrast enhancements can be used to increase the sampling depth and vice versa.

4.3.1 Biological Windows

Certain wavelength ranges in NIR with large penetration depths are known as optical or biological windows. The boundaries of these ranges are defined by water absorption, which is prominent in the NIR range (see Figure 4.1 and Figure 4.3).

The first optical window ranges between 650 nm and 950 nm. Reduced absorption allows for deeper depth penetration in tissue than in the visible region.

A second NIR optical window contains wavelengths from 1,100 nm to 1,350 nm. It is used in multiple imaging and therapeutic applications. The first and second windows are also known as therapeutic windows, as they are used in numerous therapeutic applications.

Two additional NIR optical windows were proposed recently for imaging applications. A third therapeutic spectral window with wavelengths between 1,600 nm and 1,870 nm [8] was proposed, as this range can be used for imaging more deeply into the tissue due to a reduction in scattering.

The fourth NIR window, 2,100–2,300 nm, was recently suggested for imaging biomedical applications [9]. However, using this range is in its infancy, as no suitable probes exist.

For *in vivo* fluorescence imaging, it is essential that both excitation and emission wavelengths fall within the biological windows.

4.3.2 Optical Clearing

In visible and near-infrared wavelength range, most tissues are low absorbing but highly scattering media ($\mu_s >> \mu_a$, scattering-dominated regime).

Thus, a natural approach to increase sampling depth would be to try to reduce scattering. This idea lays the foundations of the optical clearing technique, which is based on the observation that certain fluids, called optical clearing agents (OCAs), can reversibly change the light-scattering properties of tissues and blood.

In general, the scattering properties of tissue (namely, the scattering coefficient μ_s and scattering anisotropy factor g) are dependent on the refractive index mismatch between cellular tissue components: cell membrane, cytoplasm, cell nucleus, cell organelles, melanin granules, and the extracellular fluid (ECF). Thus, optical clearing is based on the decreasing refractive index mismatch between different cellular tissue components.

The root cause of the mismatch of refractive indices in tissues is the big difference in the refractive index of water, which constitutes up to 70% of the body weight, and protein and lipids. If the water has a refractive index of $n = 1.33$, proteins and lipids have a refractive index in the $n = 1.5$-1.7 range. As such, the refractive index in tissues ranges from 1.35–1.36 (extracellular fluid) to 1.360–1.375 (cytoplasm) to 1.38–1.41 (nucleus, mitochondria, and organelles) to 1.46 (cell membrane) to 1.6–1.7 (melanin) [10].

The optical clearing technique is based on the impregnation of a tissue by a biocompatible chemical agent, which decreases light scattering within the tissue. The OCAs frequently used are glucose, dextrose, fructose, glycerol, mannitol, sorbitol, propylene glycol, polypropylene glycol, and polyethylene glycol (PEG) with different molecular weights, 1,3-butanediol, 1,4-butanediol, and their combinations, and X-ray contrasting agents (verografin, trazograph, and hypaque) [11].

4.3.2.1 Mechanisms of Optical Clearing

As mentioned, scattering in biological tissues arises from spatial variations in the refractive index. From that perspective, scattering particles (organelles, protein fibrils, membranes, and protein globules) can be considered immersed into the ground substance (interstitial fluid or ECF). Due to dissolved proteins, ECF's refractive index is higher than water's and is typically assessed as $n_0 = 1.35 - 1.37$ [10].

As scattering particles (organelles, protein fibrils, membranes, and protein globules) composed of proteins and lipids, they have a greater index of refraction $\left(n_s = 1.39 - 1.47\right)$ [12] in comparison with the ground substance (ECF).

The nucleus and the cytoplasmic organelles in mammalian cells contain quite similar concentrations of proteins and nucleic acids. As such, they have refractive indices in a relatively narrow range (1.38–1.41) at a wavelength of 589 nm [10].

The refractive index of the connective-tissue fibers is approximately 1.41 and depends on the hydration of its main component, collagen [13].

Thus, the scattering in biological tissues arises from a relatively small mismatch of refractive indexes. The notable exception here is melanin, which has a much higher refractive index $(n = 1.6)$ [10].

According to Larin et al. [11], there are several primary mechanisms of light scattering reduction induced by OCAs, including 1) dehydration of tissue constituents, 2) partial replacement of the interstitial/ECF by the immersion substance, and 3) structural modification or dissociation of collagen.

The first mechanism is characteristic of highly hyperosmotic agents. It can be illustrated as follows. Suppose a hyperosmotic OCA is applied topically to the skin. Besides its diffusion into the skin, it will lead to tissue water flow outside from a tissue. It results in tissue dehydration, leading to a) matching of refractive indices of scatterers relative to the background substance and b) more effective packing of scatterers (tissue shrinkage). Both processes result in more effective light transport through the skin and reduced scattering.

The second mechanism is prevalent for fibrous tissues (the sclera, dura-mater, and dermis) as OCA's molecule sizes are much less than the mean cross-section of interfibrillar space.

4.3.2.2 Optical Clearing in Fibrous Tissues

Optical clearing has been extensively studied in fibrous or connective tissues, which include eye scleral and corneal stroma, skin dermis, cerebral membrane, muscle, vessel walls, female breast fibrous component, cartilage, tendon, etc. Intense light scattering in fibrous tissues arises from the refractive index mismatch between extracellular fluid and long fibers of scleroprotein (e.g., collagen-, elastin-, or reticulin-forming fibers) [11]. The skin represents a practically important case for optical clearing.

Optical clearing is used in tissue histology. For example, optical clearing enhances the capabilities of three-dimensional (3D) volumetric imaging of tissues, which

> provides a comprehensive view of 3D biological tissue architecture and has been used in *ex vivo* 3D skin pathology studies to investigate epidermal hyperplasia in psoriatic skin or neuronal

innervation in pruritic skin as well as dermal blood flow in *in vivo* studies [14].

In addition to dermatological applications, optical clearing can enhance OCT's capabilities for quantification of drug and molecule permeation, glucose sensing, and monitoring of skin perforation and nanoparticle delivery [11].

The excellent diffusional resistance of the skin stratum corneum (SC) makes the transdermal delivery of immersion agents and water loss by skin difficult [15]. The diffusion of water across the SC is a passive process that can be modified by applying hyperosmotic OCAs.

In addition to soft tissues, blood [16] and hard tissues like bones [17] can also be optically cleared.

4.3.2.3 Optical Clearing in Blood

The primary scatterers in the blood are red blood cells (RBCs), which in humans are anucleous cells containing approximately 70% water, 25% hemoglobin (Hb), and 5% lipids, sugars, salts, enzymes, and proteins [18]. The refractive index of human blood plasma for various wavelengths is approximately 1.33–1.35 [10]. The refractive index of dry RBCs at 550 nm ranges from 1.61 to 1.66 [19]. For physiological hemoglobin concentrations in blood (normal hematocrit, 140 g/L), its refractive index decreases from 1.369 to 1.352 as wavelength increases from 400 to 700 nm [20]. However, the hemoglobin is confined to RBCs, with a much higher concentration of 320 g/L, corresponding to the refractive index of approximately 1.42 [21]. As a result, the refractive index of the whole blood in the visible spectrum range is 1.36–1.40 [10].

Glucose is a primary OCA for optical clearing of the blood. It was found that minimal light scattering occurs at a glucose concentration in blood of 0.65 g/mL [22]. Obviously, such large concentrations of glucose can damage blood cells and vessel wall tissue. Controllable injections of small amounts of glucose in the vessel lumen can address this problem. In particular, combining endoscopic optical imaging techniques (OCT or confocal microscopy) with controllable injections can facilitate imaging of an atherosclerotic plaque with high contrast [22].

There is also the possibility of applying a small amount of free hemoglobin, which could be released owing to the local hemolysis of RBCs within the vessel area near the endoscopic optical probe as an immersion agent [23]. A 30–40% reduction of the scattering coefficient of blood in the spectral range 400–1,000 nm due to the local hemolysis (up to 20% of RBC in the close vicinity of the optical probe) was demonstrated theoretically [23] as well as experimentally [24].

4.3.2.4 Other Approaches to Optical Clearing

The squeezing (compressing) or stretching of a soft tissue causes a significant increase in its optical transmission due to reduced scattering [16]. According to Larin et al. [11], the primary mechanisms behind this phenomenon are as follows: 1) increased optical tissue homogeneity due to the removal of blood and ECF from the compressed site; 2) more close packing of tissue components leading to constructive interference (cooperative) effects; and 3) decreased tissue thickness.

4.3.3 Optimization of Illumination and Collection Geometry

As has already been mentioned, the sampling depth depends on the geometry of the data collection. Here, we split this problem into illumination and collection geometries.

4.3.3.1 Illumination Geometry

In clinical imaging settings, different illumination and data collection geometries are used. The two most common illumination geometries are wide beam diffuse illumination (e.g., ambient light) and wide beam collimated illumination (e.g., certain medical light sources, lasers).

As mentioned in section 4.1.1, the penetration depth depends on the illumination type. For example, collimated and diffuse illumination will have different penetration depths. As a result, other observables like MD or MOP are also different. For example, using the approach developed in [25] (see Appendix B for details), one can demonstrate that collimated illumination provides better image contrast than wide beam diffuse illumination.

For example, we can analyze the visualization of the upper part of the capillary loop in the dermis. In Fig.4.4A, one can see the calculated contrast for a loop assuming $\mu_a = 0.033$ mm^{-1}, $\mu_s' = 5$ mm^{-1}, and $n = 1.33$ for reticular dermis [26]), and absorption coefficient $\delta\mu_a = 28$ mm^{-1} (the whole blood with 70% oxygenation at 532 nm) and volume $V = 2x10^3 \mu$m^3 for the capillary loop.

Figure 4.4B demonstrates the contrast in the point above the defect $c(0)$ as a function of the defect depth (in mm) for wide beam diffuse (solid red line) and collimated (dotted blue line) illumination. From Figure 4.4B, one

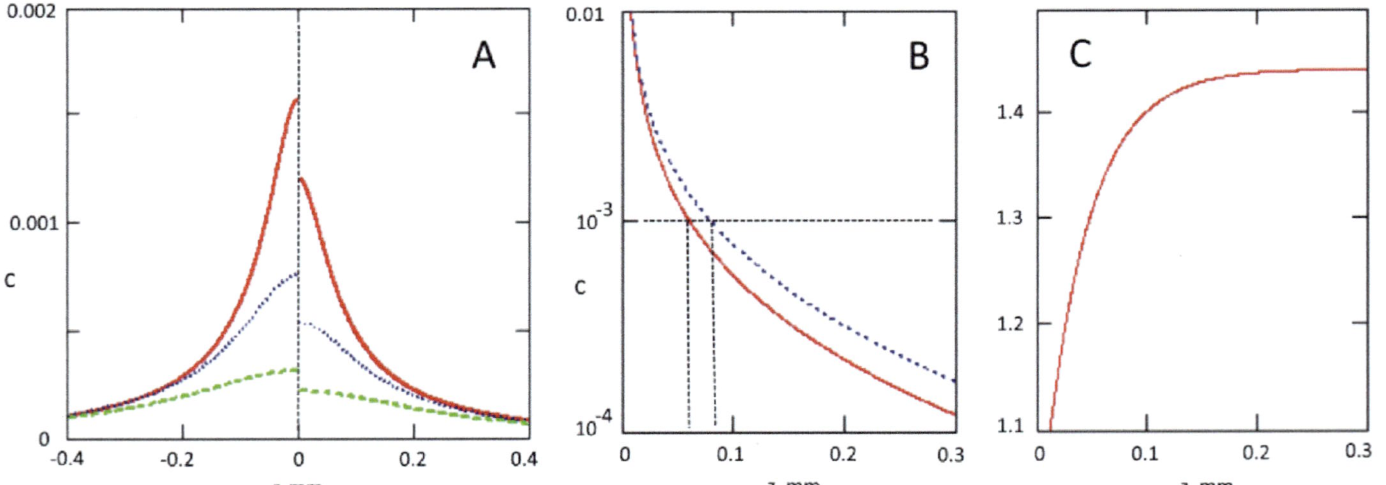

FIGURE 4.4   Panel A – radial dependence of the contrast ratio (collimated illumination – left side, diffuse illumination – right side) on the surface above a defect located at depth 0.05 mm (solid red line), 0.1 mm (dotted blue line), and 0.2 mm (dashed green line). Panel B – the contrast in the point above the defect $c(0)$ as a function of the defect depth (in mm) for wide beam diffuse (solid red line) and collimated (dotted blue line) illumination. Panel C – the ratio of contrasts for collimated illumination and diffuse illumination. Tissue parameters are: $\mu_a = 0.033$ mm^{-1}, $\mu_s' = 5$ mm^{-1}, $n = 1.33$, vessel parameters are $\delta\mu_a = 28$ mm^{-1} (the whole blood with 70% oxygenation at 532 nm) and $V = 2x10^3\,\mu m^3$.

can see that such defects can be visualized (with $c_{th} = 0.001$) till approximately 0.059 mm for diffuse illumination and 0.078 mm for collimated illumination.

Figure 4.4C demonstrates the ratio of contrasts for collimated illumination and diffuse illumination. For chosen parameters, a 40% contrast improvement can be achieved using collimated illumination. Calculations show (see Appendix B for details) that in the case of the matched boundary $(r_{21} = 0)$, the ratio of contrasts for collimated (c_c) and diffuse (c_d) illuminations can be found as

$$c_c / c_d \rightarrow 5 / \left(3 - 4\sqrt{\mu_a / 3\mu_{tr}}\right) \tag{4.10}$$

Such as, in most tissues, $\mu_s >> \mu_a$, c_c / c_d can be estimated as 5/3, which means that for the matched boundary collimated light provides 66% improvement over diffuse light.

4.3.3.2 Collection Geometry

Similarly to illumination, one could expect that the image contrast and sampling depth can be increased by discarding diffuse photons and collecting quasiballistic photons in the collection tract.

Several imaging techniques have been proposed to discard diffuse photons by narrowing the acceptance angle, including angular domain imaging [27] and numerical aperture-gated microscopy [28].

For example, angular domain imaging (ADI) uses high-aspect-ratio parallel microchannels (angular filter array or AFA) as an angular filter to capture quasiballistic photons [29]. The authors found that an angular filter with angular acceptance of about 0.4 deg yields the highest image contrast for transillumination images. The performance can be further improved by time-gating.

The use of a small numerical aperture (aperture-gated microscopy or spatio-angular filter (SAF) imaging [28]) has an advantage over ADI as it allows imaging in reflectance geometry.

Time gating (e.g., time-of-flight, ToF measurements) can also discard diffuse photons. However, while it has obvious implications for contrast enhancements (like rejecting ultrafast specular photons), increasing a sampling depth using time gating is less trivial. In particular, long quasiballistic photons can have the same delay as shorter diffuse photons. Perhaps the combination of time gating with acceptance angle gating will have the best results in this direction.

NOTE

1 It should be noted that different versions of OCT may provide functional information (like blood velocity) as well.

REFERENCES

1. JF Algorri, M Ochoa, P Roldán-Varona, et al. "Light technology for efficient and effective photodynamic therapy: A critical review". *Cancers (Basel)* 13(14): 3484 (2021).
2. S Chatterjee, PA Kyriacou. "Monte Carlo analysis of optical interactions in reflectance and transmittance finger photoplethysmography". *Sensors* 19: 789 (2019).
3. S Feng, F Zeng, B Chance. "Monte Carlo simulations of photon migration path distributions in multiple scattering media". *Proc SPIE* 1888: 78–97 (1993).
4. S Jacques, I Saidi, A Ladner, D Oelberg. "Developing an optical fiber reflectance spectrometer to monitor bilirubinemia in neonates". *Proc SPIE* 2975: 115–124 (1997).
5. G Saiko, A Douplik. "Visibility of capillaries in turbid tissues: An analytical approach". *arXiv*: 2210.04301 (2022).
6. G Saiko, A Douplik. "Reflectance of biological turbid tissues under wide area illumination: Single backward scattering approach". *Int J Photoenergy*: 241364 (2014).
7. GM Hale, MR Querry. "Optical constants of water in the 200 nm to 200 μm wavelength region". *Appl Opt* 12: 555–563 (1973).
8. LA Sordillo, S Pratavieira, Y Pu, et al. "Third therapeutic spectral window for deep tissue imaging". In *Proceedings of the SPIE 8940, Optical Biopsy XII*, p. 89400V (2014).
9. LA Sordillo, Y Pu, S Pratavieira, et al. "Deep optical imaging of tissue using the second and third near-infrared spectral windows". *J Biomed Opt* 19: 056004 (2014).
10. VV Tuchin. *Tissue Optics: Light Scattering Methods and Instruments for Medical Diagnosis*. Bellingham, WA: SPIE Press (2007).
11. KV Larin, MG Ghosn, AN Bashkatov, et al. "Optical clearing for OCT image enhancement and in-depth monitoring of molecular diffusion". *IEEE J Sel Topics Quant Electr* 18(3): 1244–1259 (2012).
12. JM Schmitt, G Kumar. "Optical scattering properties of soft tissue: A discrete particle model". *Appl Opt* 37: 2788–2797 (1998).
13. DW Leonard, KM Meek. "Refractive indices of the collagen fibrils and extrafibrillar material of the corneal stroma". *Biophys J* 72: 1382–1387 (1997).
14. Y Tan, CPL Chiam, Y Zhang, et al. "Research techniques made simple: Optical clearing and three-dimensional volumetric imaging of skin biopsies". *J Invest Derm* 140(7): 1305–1314.e1 (2020).
15. H Schaefer, TE Redelmeier. *Skin Barrier: Principles of Percutaneous Absorption*. Basel: Karger (1996).

16. VV Tuchin. *Optical Clearing of Tissues and Blood*, Vol. PM 154. Bellingham, WA: SPIE Press (2006).

17. E Genina, A Bashkatov, VV Tuchin. "Optical clearing of cranial bone". *Adv Opt Technol* 2008: 267867 (2008).

18. AN Yaroslavsky, AV Priezzhev, J Rodriguez, et al. "Optics of blood". In *Handbook of Optical Biomedical Diagnostics*, VV Tuchin, Ed. Bellingham, WA: SPIE Press, pp. 169–216 (2002).

19. G Mazarevica, T Freivalds, A Jurka. "Properties of erythrocyte light refraction in diabetic patients". *J Biomed Opt* 7: 244–247 (2002).

20. O Zhernovaya, O Sydoruk, V Tuchin, A Douplik. "The refractive index of human hemoglobin in the visible range". *Phys Med Biol* 56: 4013–4021 (2011).

21. M Friebel, M Meinke. "Model function to calculate the refractive index of native hemoglobin in the wavelength range of 250–1100 nm dependent on concentration". *Appl Opt* 45: 2838–2842 (2006).

22. VV Tuchin, X Xu, RK Wang. "Dynamic optical coherence tomography in studies of optical clearing, sedimentation, and aggregation of immersed blood". *Appl Opt* 41: 258–271 (2002).

23. V Tuchin, D Zhestkov, A Bashkatov, E Genina. "Theoretical study of immersion optical clearing of blood in vessels at local hemolysis". *Opt Exp* 12: 2966–2971 (2004).

24. G Popescu, T Ikeda, CA Best, et al. "Erythrocyte structure and dynamics quantified by Hilbert phase microscopy". *J Biomed Opt* 10: 060503 (2005).

25. G Saiko, A Douplik. "Contrast ratio during visualization of subsurface optical inhomogeneities in turbid tissues: Perturbation analysis". In *Proceedings of the 14th International Joint Conference on Biomedical Engineering Systems and Technologies (BIOSTEC 2021) – Vol. 2: BIOIMAGING*, pp. 94–102 (2021).

26. IV Meglinski, SJ Matcher. "Quantitative assessment of skin layers absorption and skin reflectance spectra simulation in the visible and near-infrared spectral regions". *Physiol Meas* 23: 741753 (2002).

27. F Vasefi, B Kaminska, GH Chapman, PKY Chan. "Angular domain imaging for tissue mapping". In *2006 Bio Micro and Nanosystems Conference*, San Francisco, CA, USA, pp. 39–45 (2006).

28. A Pandya, I Schelkanova, A Douplik. "Spatio-angular filter (SAF) imaging device for deep interrogation of scattering media". *Biomed Opt Expr* 10(9): 4656–4663 (2019).

29. F Vasefi, M Najiminaini, E Ng, et al. "Angular domain transillumination imaging optimization with an ultrafast gated camera". *J Biomed Opt* 15(6): 061710 (2010).

PART 2

Large Area Physiological Optical Imaging Techniques

PHYSIOLOGICAL OPTICAL IMAGING CAN be classified using multiple categories.

One way of thinking is that it can be skin-based vs. endoscopic. However, endoscopic techniques currently use few physiological optical imaging techniques (see Chapter 11). Thus, this separation does not provide significant methodological value.

Another approach would be to consider functional versus structural imaging. However, due to a small penetration depth, most optical modalities provide functional information. Structural imaging modalities are mostly limited to microscopic modalities, which will be considered in Chapter 12. Moreover, some structural modalities (like OCT) can also provide functional information. For example, OCT-angiography (OCTA) can provide information on choroidal vasculature and choriocapillaris blood perfusion.

The other approach would be based on a physical principle (e.g., interferometry) or field of view (macroscopic vs. microscopic). However, for example, for the field of view approach, multiple technologies (multispectral imaging or fluorescence) can be used on both scales.

Thus, to avoid duplication, we adopted the following approach. In this part of the book, we will try to categorize methods that can be used universally across macro-, micro-, and endoscopic imaging. We will categorize

DOI: 10.1201/9781003482505-6

them based on the primary underlying physical properties: absorption, scattering, fluorescence, thermal emission, etc. While this categorization has blurred lines (for example, spectroscopic techniques rely on scattering, or fluorescence requires absorption of the photon in the first place), it is still possible to find the dominant feature in most cases.

Then, in Part 3, we will consider endoscopic and microscopic techniques.

CHAPTER 5

Absorption-Based Physiological Optical Imaging Techniques

ABSORPTION-BASED TECHNIQUES REFER TO a broad range of optical techniques based on absorption spectra features of tissue chromophores. The bulk of absorption-based techniques can be categorized as spectroscopic techniques, which refer to a broad range of technologies that use spectroscopic principles to extract information about underlying tissue.

Light in the visible and near-infrared (NIR) ranges of the spectrum delivered to biological tissue undergoes multiple scattering and absorption events. Hemoglobins, water, fat, and melanin are the primary absorbers in the tissue. Thus, the transmitted or reflected light carries information about these chromophores, which can be used to assess tissue pathology.

Absorption-based techniques' primary clinical utility is providing information about tissue perfusion.[1] While it is challenging to measure perfusion noninvasively, it can be done indirectly through its proxies, i.e., blood oxygen saturation or skin temperature. For example, blood oxygen saturation in microvasculature is the balance between blood supply (perfusion) and demand (consumption). Thus, one can indirectly assess the blood perfusion by determining blood oxygen saturation in the microvasculature.

DOI: 10.1201/9781003482505-7

Absorption-based techniques do not provide direct information about perfusion. Instead, most of them provide information on the concentration (absolute or relative) of oxy- and deoxyhemoglobins. Other metrics, such as blood oxygen saturation (SO_2) and total hemoglobin (tHb), can be derived from this information. In addition, certain clinical applications analyze other chromophores (e.g., water or melanin).

For many years, most spectroscopic modalities were based on single-point measurements. For example, near-infrared spectroscopy (NIRS) typically refers to a single-point contact measurement, e.g., using a fiber probe. In recent years, the progress in cameras, image analysis techniques, and computational power has made it possible to extend the advances in biospectroscopy into clinical imaging modalities.

5.1 COLOR-BASED TECHNIQUES

Skin color is essential to skin and wound assessment, providing valuable information about physiology and pathology.

The primary utility of skin color in medicine is to serve as a proxy for skin perfusion assessment. The ability to use skin color for perfusion analysis can be attributed to the small penetration depth of the visible light in the skin (less than 2 mm), which makes the visible range of the spectrum particularly suitable for microcirculation investigations.

5.1.1 Clinical Utility

There are numerous applications of skin color analysis, which have been used across various fields of medicine for centuries.

For example, pallor is a pale skin color that can be caused by illness, emotional shock or stress, stimulant use, or anemia and is the result of a reduced amount of oxyhemoglobin. It may also be visible as pallor of the conjunctivae of the eyes on physical examination.

Similarly, skin color is an essential part of the classic definition of inflammation, which can be stated in Latin as calor, dolor, rubor, and tumor. This assonant phrase refers to the heat (calor), pain (dolor), redness (rubor), and swelling (tumor) that characterize the clinical symptoms of inflammation as they were defined in the first century AD by the Roman scholar Celsus [1].

Changes in skin color after applying pressure are an essential tool in the diagnostics of pressure injury (blanching vs. non-blanching erythema) and decreased peripheral perfusion (capillary refill time [2]), which may be indicative of cardiovascular or respiratory dysfunctions.

There are numerous attempts to develop skin color imaging modalities for diagnostics and patient monitoring. For example, Kiranantawat et al. [3] proposed an Android app for free flap monitoring. Using a k-nearest neighbor algorithm, they detected abnormal perfusion with 94%, 98%, 95%, 6%, and 1% for the sensitivity, specificity, accuracy, false-negative, and false-positive results, respectively.

5.1.2 Practical Considerations

Despite its long history and extensive medical use, visual skin color analysis has several significant drawbacks. In particular, it is a) qualitative and b) a subjective metric, c) which is not universally applicable to all skin tones (particularly for darker skin tones). Moreover, d) the color perception during physical examination is affected by ambient light intensity, color temperature (Kruithof diagram [4]), and color blindness. Finally, e) it requires a doctor's physical presence and cannot be immediately leveraged in telehealth applications.

Imaging modalities may address these drawbacks. In particular, imaging technologies may have significantly higher sensitivity than human vision. For example, the human eye can resolve contrast on a scale of 0.1. However, even consumer-grade electronics have higher contrast resolution.

The important step in skin imaging modality design is using an appropriate color representation or color scheme.[2] Multiple color spaces (models) are used for different applications such as computer graphics, image processing, TV broadcasting, and computer vision. They can be split into three groups: RGB-based color space (RGB, normalized RGB), Hue Based color space (HSI, HSV, and HSL), and Luminance based color space (YCBCr, YIQ, and YUV). Research shows that different color spaces perform differently in skin detection and analysis applications. For example, Ryu et al. [5] analyzed 13 methods for color channel combination in the color spaces currently used in photoplethysmographic imaging to determine the combination method that can improve the pulse signal quality. They found that among these methods for color channel combination, the signal extracted by the Cb +Cr combination in the YCbCr color space includes the most pulse information.

5.2 NEAR-INFRARED SPECTROSCOPY (NIRS) AND DIFFUSE OPTICAL TOMOGRAPHY (DOT)

Diffuse reflectance spectroscopy (DRS), diffuse optical spectroscopy (DOS), and near-infrared spectroscopy (NIRS) refer to a broad range of technologies that use diffuse photons to extract functional and morphological information about underlying tissue. As scattering and absorption of tissues are lower in the NIR range of spectrum, NIR light sources are often used. Thus, this group of technologies is commonly known as NIRS, particularly in neuroscience.

The utility of NIRS has been explored for multiple medical applications. For example, Poosapadari et al.[6] used spectroscopy to discriminate between the two most prevalent pathogens in DFU patients (*S. aureus* and *E. coli*), with 100% sensitivity and 75% specificity in detecting the presence of these infections and a 100% negative predicted value in excluding the infection in such wounds. In wound care, considerable differences in RHb and HbO2 were observed in ulcer regions when using DRS compared to control sites in clinical studies. In addition, the total hemoglobin (tHb) concentration was higher in foot ulcers [7].

Photoplethysmography (PPG) is a NIRS subclass that exploits the volumetric changes caused by pulse propagation. A pulse oximeter is the best-known PPG device.

The utility of NIRS is well-established in several medical fields, including patient monitoring (pulse oximeter), functional brain monitoring in neuroscience, and muscle oximetry in sports medicine.

Fundamentals of diffuse optics have been reviewed by Durduran et al. [8]. NIRS technology aims to extract the absorption coefficient of the media at several wavelengths. Then, the concentrations (absolute or relative) of chromophores (oxy-, deoxyhemoglobin, water, lipids) can be extracted from them.

NIRS methodology has three flavors: time domain, frequency domain, and continuous wave (CW).

Time-domain imaging irradiates the tissue with short (typically picoseconds) light pulses. Then, the shape of the received signal is analyzed to extract optical parameters.

In frequency-domain schema, the light source is modulated at high frequency (typically 100 MHz). Some tools have modulation at several frequencies up to 1 GHz.

Continuous-wave spectroscopy typically explores several source-to-detector distances (spatially resolved spectroscopy).

Diffuse reflectance spectroscopy (DRS), diffuse optical spectroscopy (DOS), and near-infrared spectroscopy (NIRS) typically refer to contact modalities that provide single-point measurements. NIRS modalities are usually based on the principle of spatially resolved spectroscopy.

Diffuse optical tomography (DOT) extends the NIRS or DOS to several sources and detectors. The technology relies heavily on image reconstruction. So, it is not true imaging technology but a synthetic imaging modality. DOT can be used for 2D (topographic) and 3D (tomographic) reconstructions.

5.2.1 Clinical Utility

DOS and DOT have been tried for many clinical applications. However, the primary clinical utility was established in brain imaging and mammography.

5.2.1.1 Brain Imaging

NIRS has a particularly broad range of applications in brain research. For example, it is one of the core techniques (together with functional MRI, or fMRI) of functional neuroimaging. In particular, functional NIRS (fNIRS) is an optical brain monitoring technique that uses near-infrared spectroscopy for brain function studies. Using fNIRS, brain activity is measured using near-infrared light to estimate cortical hemodynamic activity, which occurs in response to neural activity. fNIRS was cross-validated against fMRI numerous times (see, for example,[9]).

fNIRS has a particularly high potential for clinical applications, as, unlike fMRI, it allows continuous (including long-term) bedside monitoring at low cost and irrespective of age (including neonates). For example, Durduran et al. [10] tested the feasibility of longitudinal hybrid NIRS/DOS-DCS measurements on a critically ill population of acute, ischemic stroke patients.

In addition to oxy- and deoxyhemoglobin, NIRS is used to extract information on cytochrome c oxidase, which can yield crucial information about cerebral metabolism at the patient bedside, particularly for neonates [11].

5.2.1.2 Breast Cancer Imaging

Since diffuse optical tomography (DOT) showed its ability to differentiate cysts from solid tumors [12] and provide functional images, it has been tried in multiple aspects of breast cancer management, i.e., to detect breast cancer, to differentiate it, and to monitor its treatment. The utility of DOT for breast cancer imaging and clinical studies was reviewed by Grosenick et al. [13].

Early works (1990s–2000s) were devoted to developing a technology capable of detecting cancers (optical mammography). The consistent finding across different studies is that breast carcinomas were found to show a significant increase in tHb, whereas benign lesions exhibited a smaller increase. However, the specificity of optical breast imaging as a stand-alone imaging modality was found to be poor [14]. The changes in StO2 are even less consistent.

Although the original aim of developing a new tool for screening comparable to X-ray mammography has not been reached so far, optical breast imaging has found potentially new areas of application, including monitoring neoadjuvant therapy progress and determining risk populations for breast cancer development [13].

In particular, DOT has been proposed for monitoring neoadjuvant chemotherapy (NAC) [15]. NAC is increasingly used for patients with locally advanced breast cancer to shrink tumor size prior to surgery. For example, NAC has been shown to increase the rate of breast-conserving surgery [16]. The DOT utility to monitor the response to NAC is based on NIRS's capability to measure functional parameters of carcinomas at known locations within the breast. It has several advantages over traditional techniques like dynamic contrast-enhanced MRI or PET mammography, as DOT is technically less demanding and does not need a contrast agent.

5.2.2 Practical Considerations

Unlike X-ray computed tomography (CT), image reconstruction employing backprojection does not work for DOT because of diffusive light propagation. Therefore, DOT image reconstruction consists of two processes [17]: a) a forward process, which calculates light propagation and predicts measurements, and b) an optimization process that minimizes the error between the actual and predicted measurements. DOT image reconstruction is typically based on the finite element method [18]. However, other methods have emerged recently, including the boundary element method (BEM) [19]. DOT image reconstruction algorithms have been reviewed by Okawa and Hoshi [17].

DOT can be combined with other modalities (X-ray [20], MRI [21], or ultrasound [22]), which provide structural information.

Contrast can be enhanced with exogenous chromophores (like indocianine green, ICG [23]). As ICG extravasculirizes in tumor tissue, some extensions of DOT into fluorescence space (fluorescence mammography) have been proposed [24].

Limitations of diffuse optical measurements include [8] relatively short penetration beyond the skull/scalp, relatively coarse spatial resolution, limited structural/morphological information, which makes partial volume effects more difficult to account for, and difficulties determining absolute optical properties.

5.3 HYPERSPECTRAL AND MULTISPECTRAL IMAGING

Hyperspectral and multispectral imaging refer to a broad group of non-contact spectroscopy-based technologies that extract information about underlying tissues from relatively large areas.

Biomedical hyperspectral imaging (HSI) aims to record the spectrum for each image pixel and extract the concentration of tissue chromophores. Thus, hyperspectral imaging is the natural extension of color (RGB) imaging and biospectroscopy (a single-point measurement). The spectrum at each pixel as a function of a wavelength (λ, nm) can be considered a spectroscopic input, which can be decomposed, and spectral signatures can be found.

Often, the terms "hyperspectral" (HSI) and "multispectral" (MSI) imaging are used interchangeably. The subtle distinction between them is based on an arbitrary "number of bands." Multispectral imaging deals with several images at discrete and somewhat narrow bands placed at particular spectral points (e.g., isosbestic or absorption maxima). Thus, multispectral images do not produce the "spectrum" of an object but rather sample the spectrum at several points. On the other hand, hyperspectral imaging implies narrow, equally spaced spectral bands over a continuous spectral range, which can be considered a spectrum.

In biomedical applications, hyperspectral or multispectral imaging is used primarily to extract data about components of the blood, which are chromophores in the visible and NIR spectrum [25]. The primary outcome of HSI/MSI methods is tissue oxygen saturation (SO_2) maps, which indicate abnormalities in blood circulation. Occasionally, oxyhemoglobin (HbO_2), deoxyhemoglobin (RHb), and total hemoglobin (tHb) or their proxies are also reported.

HSI/MSI is increasingly being used within different clinical diagnostic areas. In line with many reported clinical applications [25], hyperspectral imaging demonstrated its utility in wound care in general and diabetic foot ulcers in particular. For example, hyperspectral imaging has been used to assess tissue viability and health in diabetes patients at risk of foot ulceration [26, 27].

5.3.1 Clinical Utility

The primary clinical utility of HSI/MSI techniques is to determine blood circulation abnormalities in tissues (perfusion).

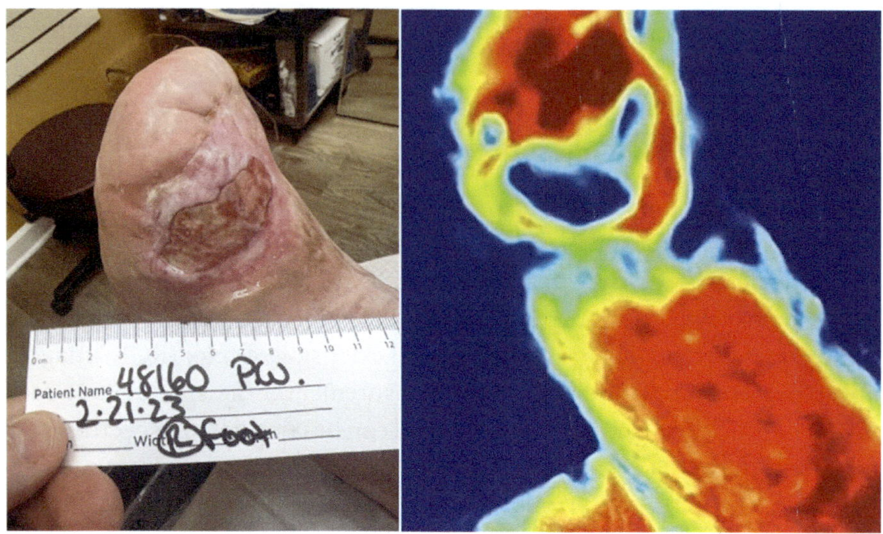

FIGURE 5.1   HSI oxygen content imaging. A diabetic foot ulcer was assessed using HSI imaging (right panel). A false color of the image shows low oxygen content (blue and green hues) in the wound bed compared to the surrounding tissue (red). Interestingly, the peri-wound area exhibits areas of increased oxygen content, probably due to local hyperemia to the region (asterisk), and areas of lower oxygen content where skin breakdown is visible (arrows). (Image courtesy of Dr. Matthew Regulsky).

5.3.1.1 Oncology

The applications of HSI in tumor detection can be categorized into three groups [28]: *ex vivo* imaging for pathological samples, *in vivo* imaging on shallow surfaces, and intraoperative applications.

5.3.1.1.1 *Ex Vivo* Imaging for Pathological Samples   Hyperspectral imaging, as a technique for assessment of resection margins during surgery, can be performed both *in vivo* and *ex vivo*. *Ex vivo* applications are less demanding. However, they still can provide valuable information within minutes. In addition, HSI/MSI allows label-free imaging without sample preparation. Thus, it can be used in conjunction with established histological techniques.

For example, Lu et al. [29] used HSI to analyze fresh surgical specimens from patients ($n = 36$) with head and neck cancers undergoing surgical resection, aiming to detect and delineate cancerous regions validated by histopathologic diagnosis. They achieved a 91% sensitivity and

91% specificity, which were more accurate than autofluorescence imaging (p < 0.05) or fluorescence imaging of 2-NBDG (p < 0.05) and proflavine (p < 0.05).

5.3.1.1.2 *In Vivo* Imaging on Shallow Surfaces While unable to substitute traditional biopsy, HSI/MSI holds great promise as an optical biopsy technique for screening purposes. It has been tried for multiple types of cancers. Oftentimes, HSI/MSI is combined with other techniques like fluorescence imaging or orthogonal polarized reflectance imaging (see section 5.5.1). For example, Roblyer et al. [30] reported using a multispectral digital microscope (DMD) to detect oral neoplasia in a pilot clinical trial.

5.3.1.1.3 Intraoperative Applications HSI/MSI can be used for surgery planning and intraoperative guidance. In particular, HSI/MSI can be used to a) visualize the surgical environment, b) intraoperative navigation for minimal injury, and c) delineate tumor margins.

For example, during visualization of the surgical environment, hypoxic areas can be identified.

In intraoperative navigation to minimize injury, HSI/MSI can be used to differentiate various types of tissues. For example, Zuzak et al. [31] used a NIR HSI system combined with a conventional laparoscopy to show a real-time, *in vivo* display of tissues, status, and anatomical structures during a cholecystectomy. The results showed that even the gallbladder could be differentiated from the liver according to the chemical composition of different tissues.

In addition to *ex vivo* tumor margin analysis, HSI/MSI shows promise to identify tumor margins *in vivo*. As an initial step in this direction, Fabelo et al. [32] created a hyperspectral database of *in-vivo* human brain tissues. In particular, they acquired 36 hyperspectral images from 22 different patients. More than 300,000 spectral signatures were labeled from these data using a semiautomatic methodology based on the spectral angle mapper algorithm. Tissues were classified into four classes: normal tissue, tumor tissue, blood vessel, and background elements.

5.3.1.2 Wound Care

The applications of HSI/MSI in wound care were reviewed in [33]. In wound care, HSI/MSI demonstrated its potential to provide information on wound healing and predict diabetic foot ulcer (DFU) development.

5.3.1.2.1 Blood Oxygen Saturation and Wound Healing   Khaodhiar et al. [26] measured RHb and HbO_2 in nondiabetic controls (n=14) as well as diabetic patients with (n=10) and without DFU (n=13) four times over six months using HSI. HbO_2 and RHb measurements in the peri-wound of nonhealing ulcers were lower than in healing ulcers (p< 0.01) and contra-lateral foot (p< 0.001).

Nouvong et al. [27] measured HbO_2 and RHb in diabetic patients (n=44) over 24 weeks using HSI. They found that in the peri-wound area, HbO_2 and SO2 were higher around DFUs that healed ($85\pm21$and $66\pm9\%$, respectively) vs. nonhealing ($64\pm22$and $60\pm10\%$, respectively).

Jeffcoate et al. [34] compared SO_2 measurements using HSI against blood gas analysis in blood samples (in vitro) in patients with DFUs (n=43). They found a negative association between $SO_2$and healing by 12 weeks (p = 0.009) and a significant positive correlation between oxygenation assessed by HSI and time to healing (p = 0.03). In addition, a strong cor-relation between SO_2 and blood gas measurements (r = 0.994) was found.

A retrospective clinical study was conducted to determine if MSI/HSI technology can evaluate wounds and adjacent soft tissues, identify pat-terns involved in tissue oxygenation and wound healing, and predict which wounds may or may not heal. Most wounds progressing toward healing showed a wispy and gradual decrease in SO_2 as the distance from the wound increased. Landsman [35] described this as resembling a ray of sunlight, with the central red region moving to a combination of yellow and red as the distance from the wound increased. This was probably the most predictive sign that the wound was starting to heal. An example of a positive response to hyperbaric oxygen therapy in a patient with arterial insufficiency is depicted in Figure 5.2.

5.3.1.2.2 Prediction of DFU Development   Yudovsky et al. [37] assessed the utility of HSI in predicting DFU development. Diabetic patients ($n = 54$) at risk of DFU were monitored and evaluated retrospectively for ulceration. They found that the difference in HbO_2 (and RHb at a minor degree) value between wound and peri-wound was a predictor of ulceration. The authors constructed an index that could predict tissue at risk of ulceration with a sensitivity and specificity of 95 and 80%, respectively, for images taken, on average, 58 days before skin breakdown.

5.3.1.3 Other Clinical Applications

Yudovsky et al. [38] demonstrated the utility of biospectroscopy and HSI/MSI in determining the thickness of the epidermis layer.

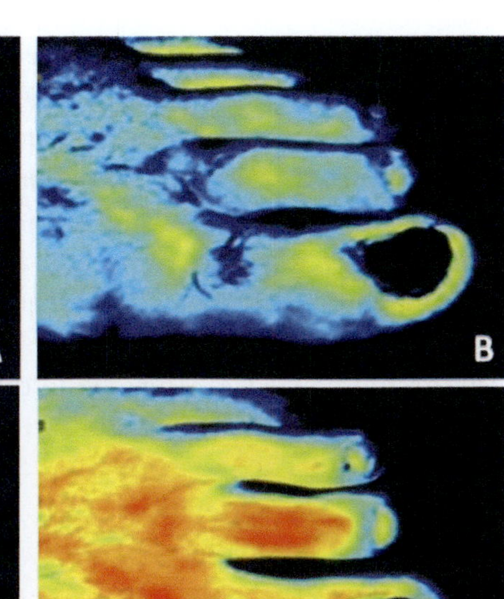

FIGURE 5.2   A positive response to hyperbaric oxygen therapy in a patient with arterial insufficiency. (A) Baseline clinical photograph; and (B) near-infrared image assessment of the left dorsal foot in a patient with arterial insufficiency referred by hyperbaric oxygen therapy. Note reduced signal to the dorsal foot (C) immediately prior to a single session of hyperbaric oxygen therapy and immediately following (D) this treatment. (Reproduced from [36] under CC BY 4.0 license).

Several studies demonstrated the utility of HSI/MSI to detect tissue water content, which can be helpful for several clinical indications, including deep tissue injury and preclinical edema identification. It can also be relevant to the diabetic population. For example, in [39], the authors found differences in skin water content between diabetic and nondiabetic populations. In [40], the authors used the 1,400–1,500 nm band to extract the skin's hydration. In [41], this approach was extended to the 970 nm range, which allows deeper light penetration into the tissue.

5.3.2 HSI/MSI: Practical Considerations

There are two different computational approaches to HSI/MSI imaging. The first one (mostly common for MSI systems) uses spectral information to extract concentrations of chromophores explicitly. The second one (which is getting increasingly popular across HSI-based systems) is combining signals from multiple (or all) spectral bands into a classi-fier, differentiating between healthy and pathological tissue. Typically, it includes training on a dataset (supervised machine learning or ML). More recent approaches in this direction include feeding the hyperspectral imaging data into neural networks (e.g., convolutional neural networks or CNN). For example, Halicek et al. [42] combined HSI with CNNs to perform an optical biopsy of *ex-vivo*, surgical gross-tissue specimens col-lected from 21 patients undergoing surgical cancer resection of squamous cell carcinoma (SCCa). They found that CNN can distinguish SCCa from normal aerodigestive tract tissues with an area under the receiver operator curve (AUC) of 0.82.

There is growing evidence that the epidermal layer can impact the extraction of physiological parameters in a number of ways.

It is known that skin tone plays an important role. In particular, it is expected that HSI/MSI can extract physiological parameters only from medium and lightly pigmented skin (skin tone I-IV).

Epidermal thickness also plays an important role. In particular, it has been shown [43] that the dermis almost does not contribute to reflected signal for stratum corneum thickness of 1.5 mm and more. Thus, inter-preting physiological parameters extracted from areas with thick epider-mal layers, e.g., calluses, should be taken cautiously.

Because various methods, protocols, and study populations are being used, the literature often reports controversial measurements and conclusions.

Several critical factors may impact study results. One can be attributed to the HSI wavelength range. Currently, there are two different HSI/MSI approaches to measuring oxyhemoglobin, which are based on the incident light range. One method uses light in the visible spectrum. The other is to use light in the near-infrared (NIR) range. The longer NIR wavelength provides deeper penetration, which may be seen as a benefit until one con-siders the vascular anatomy (see Figure 5.3). The visible light method only inspects the most superficial papillary blood supply, which supplies the wound margins. In other words, it is a very discrete measurement of a single anatomic layer.

In contrast, the NIR approach measures two layers: the superficial pap-illary layer and the deep reticular layer. These two layers have different functions, and the combined measurement of oxyhemoglobin in two lay-ers is more difficult to interpret. This may partially explain the variability

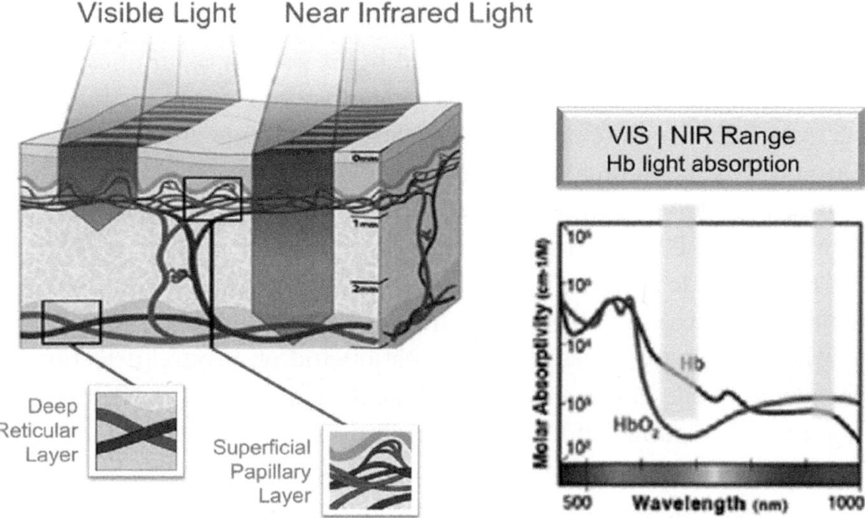

FIGURE 5.3 Oxyhemoglobin and deoxyhemoglobin wavelength absorption curves can be distinguished in visible and near-infrared ranges, as highlighted in grey. Visible light spectroscopy measures only the superficial papillary layer associated with healing granulation tissue. NIR spectroscopy penetrates deeper and is a combined measurement of the reticular and papillary systems. (Modified from [44] with permissions).

of NIR-based measurements and the interpretation of these measurements as healing predictors (see Figure 5.3).

Another important shortcoming of various HSI/MSI studies is that they often report HbO_2 and RHb in some arbitrary units, which don't allow direct interpretation or comparison. Potentially, it can be improved using a so-called "physiological" calibration. In this case, a calibration image is taken on the patient's unaffected skin or mucosa area (e.g., eyelid).

While typically, HSI/MSI is used to obtain static information, it can be used to visualize dynamic processes as well. For example, Bouchard et al. [45] developed multispectral optical intrinsic signal imaging (MS OISI) to analyze cortical hemodynamic responses in the exposed brain to stimulation. MS OISI data images are acquired at interleaved blue (470 nm) and green (530 nm) illuminations to map changing oxy, deoxy- and total hemoglobin concentrations (HbO_2, Hb, and tHb) at high speed (220 fps).

5.4 PHOTOACOUSTICS

If the absorbed energy is not re-emitted in the process of luminescence, it will be transferred to other systems, which may give rise to the local temperature. The rise in temperature can lead to multiple phenomena, from sound wave generation to ultimate tissue destruction.

It gives rise to two closely related technologies: photoacoustic (PA) and photothermal (PT) imaging.

Photoacoustic imaging is based on sound wave generation caused by a rapid tissue expansion due to local heating by the pulsed laser light. It is a hybrid technology. It uses optical excitation and ultrasound detection. In particular, it relies on well-established medical ultrasound technology that captures ultrasound (US) waves using conventional US transducers and imaging using the US image generation procedure.

Photothermal imaging is based on changing the tissue's optical properties (e.g., refractive index) caused by the pulsed laser light's local tissue heating. Unlike photoacoustic imaging, it is an all-optical technology.

However, these technologies are on different tracks to clinical translation. While photoacoustic imaging is well researched, and multiple groups work on its clinical translations in different medical fields, photothermal imaging is still a green-field modality. This chapter will consider photoacoustic imaging as a more established medical technology that has been studied in multiple clinical studies. Photothermal imaging is a microscopic technique. Thus, it is more relevant to microscopic techniques and will be considered in Chapter 12 (Microscopy) (see section 12.2.3).

Photoacoustic imaging (PAI), Photoacoustic tomography (PAT), or optoacoustic (OA) tomography uses pulsed laser light to irradiate tissues. As a result, pressure waves are produced due to the increased temperature and volume [46]. The process of photoacoustic signal generation can be described in three steps: (1) an object absorbs light, (2) the absorbed optical energy is converted into heat and generates a temperature rise, and (3) thermoelastic expansion takes place, resulting in the emission of acoustic waves [47]. For effective PAT signal generation, the laser pulse duration should be less than the thermal and stress confinement times [48]. Thus, the typical laser pulse duration used in PAT is on the scale of several nanoseconds or less.

A high-frequency ultrasound transducer monitors these pressure waves, and a 3D reconstruction is performed. Visible and near-infrared light is used to perform PAT. The contrast in PAT images depends mainly upon the absorption properties of tissues [49]. The optical absorption is specific to tissue chromophores, and by changing the wavelength of light, one can tune the images to obtain enhanced contrast for particular chromophores.

While PA can be based on exogenous or endogenous chromophores, most PA applications are based on blood chromophores (hemoglobins). By projecting light at hemoglobin's peak absorption, inflamed hyperemic tissue appears dark (hypoechoic) on PAT, while surrounding tissues reflecting such light waves appear bright (hyperechoic).

5.4.1 Clinical Utility

Lee et al. [50] have reviewed the clinical applications of PAI. They identified that the main clinical utility of PAI is to provide complementary molecular functional information of the region, augmenting the morphological information from ultrasound (US). However, photoacoustic imaging has also demonstrated its utility in providing novel, clinically viable information.

5.4.1.1 Oncology

PA has demonstrated its utility in augmenting traditional US imaging for thyroid, breast, and prostate cancers.

In thyroid cancers, PA can extract metabolic information, thus enhancing the current diagnostic method for thyroid nodules (US-based risk stratification) that relies solely on the shape of the nodules, which is characterized by low specificity and unnecessary biopsies. In particular, Roll et al. [51] used a multispectral PA system to distinguish between benign and malignant nodules. They acquired images from Graves' disease lobes ($n = 6$), healthy lobes ($n = 8$), benign lobes ($n = 13$), and malignant lobes ($n = 3$). They found that malignant nodules, compared to benign nodules, demonstrated lower SO_2 levels ($p = 0.0393$) and fat content ($p = 0.1295$).

In breast cancer, Neuschler et al. [52] conducted a multi-site study with 2,105 patients to test the diagnostic utility of multispectral PAI vs. standard US for breast cancer diagnosis. They found that PA exceeds US in specificity by 14.9% ($p < 0.0001$). In particular, PA/US downgraded 40.8% of benign mass reads, with a specificity of 43.0% for PA/US versus 28.1% for the internal US of the PA/US device. Sensitivity for biopsied malignant masses was 96.0% for PA/US and 98.6% for US ($p < 0.0001$).

In prostate cancer, Horiguchi et al. [53] conducted a pilot study to explore the potential of transrectal PAUSI in visualizing microvasculature in periprostatic tissues from prostate cancer patients ($n = 7$). In particular, during radical prostatectomies, PA and US images of the periprostatic tissues were acquired to detect the location and extent of neurovascular bundles (NVBs).

Kothapalli et al. [54] used PAUSI combined with injection of indocyanine green (ICG) to acquire high-contrast and high-resolution PA images through transrectal PAI, facilitating the visualization of prostate vascularity and surrounding tissues.

In novel applications, PA was used for noninvasive melanoma staging and delineating cancerous tissue boundaries. For example, Kim et al. conducted a study demonstrating the feasibility of multispectral PAI in distinguishing melanoma from normal tissue *ex vivo* [55] and *in vivo* [56]. The resulting spectrally unmixed PA images exhibited remarkable accuracy in depicting melanoma contours by visualizing the concentration of melanin. Furthermore, the PA images effectively measured the melanoma depth, demonstrating a mean absolute error of 0.36 mm compared to histopathological results.

5.4.1.2 Vascular Assessments

PA has demonstrated its utility in augmenting traditional US imaging for investigating carotid artery pathologies. In particular, Ivankovic et al. [57] used a handheld 3D multispectral PAI system for monitoring the bifurcation area of the human carotid artery, particularly in relation to ischemic stroke. A higher contrast-to-noise ratio was observed with wavelengths in the range of 800–900 nm (18.8 ±4.28 dB) compared to shorter wavelengths of 730–760 nm (9.18 ±3.8 dB), which is in line with high blood oxygen saturation in the arterial compartment.

In novel applications, PA was used for peripheral vascular investigations, as well as muscular and gastrointestinal studies.

Yang et al. [58] used PA to monitor vascular dysfunction in diabetic patients. They imaged the dorsalis pedis artery in the distal hallux region of diabetic patients ($n = 7$) and healthy volunteers ($n = 7$). By applying venous and arterial occlusion, they found a noticeable time delay in recovery in the arteries of the diabetic group (healthy: 14.66 ± 1.10, diabetic: 24.05 ± 5.36 s).

Yang et al. [59] used PAI to character circulatory and metabolic functions by monitoring hemodynamic changes in the skeletal muscle (forearm) during cuff occlusion. They found a similar trend to that of conventional NIRS.

Knieling et al.[60] presented a preliminary study on the application of PAI for assessing Crohn's disease based on the correlation between hemoglobin content and intestinal inflammation. Multispectral PA images were acquired from 108 patients. The PA responses were compared between active and nonactive patients, as determined by the basis of clinical scoring (Harvey–Bradshaw index), endoscopic scoring (Simplified Endoscopic Score for Crohn's disease), and histologic scoring (modified Riley score) as reference tests. The initial results showed significant differences between active and nonactive diseases for all the MSOT values except for oxygen saturation ($p < 0.0001$).

5.4.2 Practical Considerations

More commonly, PA is employed by dual-modal PA and ultrasound imaging (PAUSI) systems. Oftentimes, researchers use multispectral systems with a typical wavelength range of 700–1190 nm, which allows for visualizing and separating chromophores of interest (hemoglobins, lipids, water, and melanin).

The most commonly used PAI light source is Nd: YAG laser with pulse repeat frequency 10–100 Hz.

Reducing the imaging depth in PA imaging improves spatial resolution and vice versa. The imaging depth ratio with PAT's spatial resolution is typically ~ 200 [61]. The axial resolution also depends on a transducer's geometrical focus and bandwidth [62].

5.5 OTHER TECHNIQUES

Multiple other techniques have emerged. Shortly, we present a nonexhaustive list of technologies that were tried in clinical studies or have already been translated into clinical practice.

5.5.1 Orthogonal Polarization Spectral Imaging

Orthogonal polarization spectral imaging (OPS) is based on illuminating skin tissue with linearly polarized light and collecting resultant depolarized photons scattered by the tissue components using a polarizer positioned orthogonal to the plane of illuminating light [63]. It increases sensitivity by discarding single scattered photons, mostly from specular (Fresnel) reflection on surfaces. In OPS imaging, the microvasculature is illuminated with polarized green light. The technique can be used to visualize vessels and red blood cells (which absorb light in the green range of the spectrum). Thus, it can be particularly helpful in dermatology and rheumatology. The method has been mainly applied to detect burn depth [64]. However, it was also used on wounds with other etiologies (e.g., chronic venous insufficiency [65]).

5.5.2 Perfusion Imaging

Several groups recently proposed different versions of perfusion imaging. This group of technologies is based on remote photoplethysmography (rPPG), also called imaging photoplethysmography (iPPG) [66] or video photoplethysmography. It creates a 2D map of the amplitude of blood pulsations in a target area. The technology is based on the observations that the blood pulsations caused by pulse propagation vary in amplitude depending on multiple physiological parameters, including local hydrostatic pressure.

5.5.2.1 Clinical Utility

Tissue oximetry is an application of remote photoplethysmography with the largest clinical potential. This technology is particularly promising for the monitoring of neonates. Similarly to traditional (contact) pulse oximetry, the rPPG-based oximetry is based on two wavelengths (typically in red and near-infrared spectrum ranges), which provide significant contrast between oxy- and deoxyhemoglobins. The feasibility of camera-based contactless pulse oximetry has been demonstrated theoretically [67] and in small feasibility studies [68]. In particular, Verkruysse et al. [68] demonstrated that error stemming from variation among individuals is significantly lower (<1.65%) than that required by the International Organization for Standardization standard (<4%).

In addition, perfusion imaging can be used to identify low-perfused areas (e.g., occlusions) [69] and monitor reperfusion intraoperatively. In particular, it was used for reperfusion monitoring during abdominal surgery [65] and free microvascular anastomosed fasciocutaneous flaps [70].

5.5.2.2 Practical Considerations

The contrast and relative amplitude of the rPPG signal are quite low (<1%). Several techniques discussed in Chapter 3 are used to increase the contrast of perfusion imaging, including narrow band imaging, polarization gating, lock-in, and ensemble binning. Another shortcoming of the technology is susceptibility to motion artifacts.

5.6 CONCLUSIONS

Absorption-based techniques refer to a broad range of optical techniques based on absorption spectra features of tissue chromophores. They range from simple and well-established techniques used for centuries (e.g., skin color assessment during physical examination) to sophisticated novel ones like photoacoustic imaging. Hemoglobins, water, fat, and melanin are the primary absorbers in the tissue.

Multiple absorption-based techniques have been developed for research and clinical purposes. Absorption-based techniques' primary clinical utility is providing information about tissue perfusion. Absorption-based methods assess tissue perfusion indirectly (for example, blood oxygenation).

NOTES

1 Perfusion refers to fluid movement through the circulatory or lymphatic system towards a particular organ or tissue, typically in reference to the blood supply to capillary beds within tissues. The measurement of perfusion is based on the amount of blood delivered to the tissue per unit time per unit tissue mass ([ml/ml/s] or [ml/100 g/min]).
2 Color space is a mathematical model to represent color information as three or four different color components.

REFERENCES

1. L Ciaccia. "Fundamentals of inflammation". *Yale J Biol Med* 84(1): 64–65 (2011).
2. Y Monteerarat, R Limthongthang, P Laohaprasitiporn, T Vathana. "Reliability of capillary refill time for evaluation of tissue perfusion in simulated vascular occluded limbs". *Eur J Trauma Emerg Surg* 48(2): 1231–1237 (2022).
3. K Kiranantawat, N Sitpahul, P Taeprasartsit, et al. "The first Smartphone application for microsurgery monitoring: SilpaRamanitor". *Plast Reconstr Surg* 134(1): 130–139 (2014).
4. Kruithof Curve. In *Wikipedia*. https://en.wikipedia.org/wiki/Kruithof_curve (accessed May 24, 2023)
5. J Ryu, S Hong, S Liang, et al. "Research on the combination of color channels in heart rate measurement based on photoplethysmography imaging". *J Biomed Opt* 26(2): 025003 (2021).
6. S Poosapadi Arjunan, AN Tint, B Aliahmad, et al. "High-resolution spectral analysis accurately identifies the bacterial signature in infected chronic foot ulcers in people with diabetes". *Int J Low Extrem Wounds* 17(2): 78–86 (2018).
7. S Anand, N Sujatha, VB Narayanamurthy, et al. "Diffuse reflectance spectroscopy for monitoring diabetic foot ulcer – a pilot study". *Opt Lasers in Eng* 53: 1–5 (2014).

8. T Durduran, R Choe, WB Baker, AG Yodh. "Diffuse optics for tissue monitoring and tomography". *Rep Prog Phys* 73(7): 076701 (2010).

9. TJ Huppert, RD Hoge, SG Diamond, et al. "A temporal comparison of BOLD, ASL, and NIRS hemodynamic responses to motor stimuli in adult humans". *Neuroimage* 29(2): 368–382 (2006).

10. T Durduran, C Zhou, BL Edlow, et al. "Transcranial optical monitoring of cerebrovascular hemodynamics in acute stroke patients". *Opt Express* 17(5): 3884–3902 (2009).

11. G Bale, CE Elwell, I Tachtsidis. "From Jöbsis to the present day: A review of clinical near infrared spectroscopy measurements of cerebral cytochrome-c-oxidase". *J Biomed Opt* 21(9): 091307 (2016).

12. X Gu, Q Zhang, M Bartlett, et al. "Differentiation of cysts from solid tumors in the breast with diffuse optical tomography". *Acad Radiol* 11(1): 53–60 (2004).

13. D Grosenick, H Rinneberg, R Cubeddu, P Taroni. "Review of optical breast imaging and spectroscopy". *J Biomed Opt* 21(9): 091311 (2016).

14. D Floery, TH Helbich, CC Riedl, et al. "Characterization of benign and malignant breast lesions with computed tomography laser mammography (CTLM): initial experience". *Invest Radiol* 40(6): 328–335 (2005).

15. DB Jakubowski, AE Cerussi, FP Bevilacqua, et al. "Monitoring neoadjuvant chemotherapy in breast cancer using quantitative diffuse optical spectroscopy: a case study". *J Biomed Opt* 9(1): 230–238 (2004).

16. SP Gampenrieder, G Rinnerthaler, R Greil. "Neoadjuvant chemotherapy and targeted therapy in breast cancer: Past, present, and future". *J Oncol* 2013: 732047 (2013).

17. S Okawa, Y Hoshi. "A review of image reconstruction algorithms for diffuse optical tomography". *Appl Sci* 13: 5016 (2023).

18. H Dehghani, ME Eames, PK Yalavarthy, et al. "Near infrared optical tomography using NIRFAST: Algorithm for numerical model and image reconstruction". *Commun Numer Methods Eng* 25: 711–732 (2008).

19. MA Ansari, S Alikhani, E Mohajerani, R Massudi. "The numerical and experimental study of photon diffusion inside biological tissue using boundary integral method". *Opt Commun* 285(5): 851–855 (2012).

20. Q Fang, SA Carp, J Selb, et al. "Combined optical imaging and mammography of the healthy breast: Optical contrast derived from breast structure and compression". *IEEE Trans Med Imag* 28(1): 30–42 (2008).

21. V Ntziachristos, XH Ma. "Time-correlated single photon counting imager for simultaneous magnetic resonance and near-infrared mammography". *Rev Sci Instrum* 69(12): 4221 (1998).

22. Q Zhu. "Optical tomography with ultrasound localization: Initial clinical results and technical challenges". *Technol Cancer Res Treat* 4(3): 235–244 (2005).

23. V Ntziachristos, AG Yodh, M Schnall, B Chance. "Concurrent MRI and diffuse optical tomography of breast after indocyanine green enhancement". *PNAS* 97(6): 2767–2772 (2000).

24. A Corlu, R Choe, T Durduran, et al. "Three-dimensional in vivo fluorescence diffuse optical tomography of breast cancer in humans". *Opt Express* 15(11): 6696–6716 (2007).

25. G Lu, B Fei. "Medical hyperspectral imaging: A review". *J Biomed Opt* 19: 010901 (2014).

26. L Khaodhiar, T Dinh, KT Schomacker, et al. "The use of medical hyperspectral technology to evaluate microcirculatory chaves in diabetic foot ulcers and to predict clinical outcomes". *Diab Care* 30(4): 903–910 (2007).

27. A Nouvong, B Hoogwerf, E Mohler, et al. "Evaluation of diabetic foot ulcer healing with hyperspectral imaging of oxyhemoglobin and deoxyhemoglobin". *Diab Care* 32(11): 2056–2061 (2009).

28. M Halicek, H Fabelo, S Ortega, et al. "In-vivo and ex-vivo tissue analysis through hyperspectral imaging techniques: Revealing the invisible features of cancer". *Cancers (Basel)* 11: 756 (2019).

29. G Lu, JV Little, X Wang, et al. "Detection of head and neck cancer in surgical specimens using quantitative hyperspectral imaging". *Clin Cancer Res* 23: 5426–5436 (2017).

30. D Roblyer, R Richards-Kortum, K Sokolov, et al. "Multispectral optical imaging device for in vivo detection of oral neoplasia". *J Biomed Opt* 13(2): 024019 (2008).

31. KJ Zuzak, SC Naik, G Alexandrakis, et al. "Characterization of a near-infrared laparoscopic hyperspectral imaging system for minimally invasive surgery". *Anal Chem* 79: 4709–4715 (2007).

32. H Fabelo, S Ortega, A Szolna, et al. "In-vivo hyperspectral human brain image database for brain cancer detection". *IEEE Access* 7: 39098–39116 (2019).

33. G Saiko, P Lombardi, Y Au, et al. "Hyperspectral imaging in wound care: A systematic review". *Int Wound J* 17(6): 1840–1856 (2020).

34. WJ Jeffcoate, DJ Clark, N Savic, et al. "Use of HSI to measure oxygen saturation in the lower limb and its correlation with healing of foot ulcers in diabetes". *Diab Med* 32(6): 798–802 (2015).

35. A Landsman. "Visualization of wound healing progression with near infrared spectroscopy: A retrospective study". *Wounds* 32(10): 265–271 (2020).

36. J Arnold, VL Marmolejo. "Interpretation of near-infrared imaging in acute and chronic wound care". *Diagnostics* 11: 778 (2021).

37. D Yudovsky, A Nouvong, K Schomacker, L Pilon. "Assessing diabetic foot ulcer development risk with hyperspectral tissue oximetry". *J Biomed Opt* 16(2): 026009 (2011).

38. D Yudovsky, L Pilon. "Rapid and accurate estimation of blood saturation, melanin content, and epidermis thickness from spectral diffuse reflectance". *Appl Opt* 49(10): 1707–1719 (2010).

39. HN Mayrovitz, A McClymont, N Pandya. "Skin tissue water assessed via tissue dielectric constant measurements in persons with and without diabetes mellitus". *Diab Tech Theur* 15(1): 1–6 (2013).

40. M Attas, T Posthumus, B Schattka, et al. "Long-wavelength near-infrared spectroscopic imaging for in-vivo skin hydration measurements". *Vib Spectrosc* 28(1): 37–43 (2002).

41. G Saiko. "On the feasibility of skin water content imaging adjuvant to tissue oximetry". *Adv Exp Med Biol* 1269: 191–195 (2021).

42. M Halicek, JV Little, X Wang, et al. "Optical biopsy of head and neck cancer using hyperspectral imaging and convolutional neural networks". *J Biomed Opt* 24(3): 1–9 (2019).

43. G Saiko. "Callus thickness determination adjuvant to tissue oximetry imaging". In *Proceedings of the 15th International Joint Conference Biomedical Engineering Systems and Technologies (BIOSTEC 2022) – Vol. 2: BIOIMAGING*, pp. 147–152 (2022).

44. C Weinkauf, A Mazhar, K Vaishnav, et al. "Near-instant noninvasive optical imaging of tissue perfusion for vascular assessment". *J Vasc Surg* 69(2): 555–562 (2019).

45. MB Bouchard, BR Chen, SA Burgess, EM Hillman. "Ultra-fast multispectral optical imaging of cortical oxygenation, blood flow, and intracellular calcium dynamics". *Opt Express* 17(18): 15670–15678 (2009).

46. MJ Leahy, JG Enfield, NT Clancy, et al. "Biophotonic methods in microcirculation imaging". *Med Laser Appl* 22: 105–126 (2007).

47. J Xia, J Yao, LV Wang. "Photoacoustic tomography: Principles and advances". *Electromagn Waves (Camb)* 147: 1–22 (2014).

48. LV Wang. "Tutorial on photoacoustic microscopy and computed tomography". *IEEE J Sel Top Quantum Electron* 14: 171–179 (2008).

49. P Beard. "Biomedical photoacoustic imaging". *Interface Focus* 1: 602–631 (2011).

50. H Lee, S Han, H Kye, et al. "A review on the roles of photoacoustic imaging for conventional and novel clinical diagnostic applications". *Photonics* 10: 904 (2023).

51. W Roll, NA Markwardt, M Masthoff, et al. "Multispectral optoacoustic tomography of benign and malignant thyroid disorders: A pilot study". *J Nucl Med* 60: 1461–1466 (2019).

52. EI Neuschler, R Butler, CA Young, et al. "A pivotal study of optoacoustic imaging to diagnose benign and malignant breast masses: A new evaluation tool for radiologists". *Radiology* 287: 398–412 (2017).

53. A Horiguchi, K Tsujita, K Irisawa, et al. "A pilot study of photoacoustic imaging system for improved real-time visualization of neurovascular bundle during radical prostatectomy". *Prostate* 76: 307–315 (2016).

54. S-R Kothapalli, GA Sonn, JW Choe, et al. "Simultaneous transrectal ultrasound and photoacoustic human prostate imaging". *Sci Transl Med* 11: eaav2169 (2019).

55. J Kim, Y Kim, B Park, et al. "Multispectral ex vivo photoacoustic imaging of cutaneous melanoma for better selection of the excision margin". *Brit J Dermatol* 179: 780–782 (2018).

56. B Park, C Bang, C Lee, et al. "3D wide-field multispectral photoacoustic imaging of human melanomas in vivo: A pilot study". *J Eur Acad Dermatol* 35: 669–676 (2020).

57. I Ivankovic, E Mercep, C-G Schmedt, et al. "Real-time volumetric assessment of the human carotid artery: Handheld multispectral optoacoustic tomography". *Radiology* 291: 45–50 (2019).

58. J Yang, G Zhang, Q Shang, et al. "Detecting hemodynamic changes in the foot vessels of diabetic patients by photoacoustic tomography". *J Biophoton* 13: e202000011 (2020).

59. J Yang, G Zhang, W Chang, et al. "Photoacoustic imaging of hemodynamic changes in forearm skeletal muscle during cuff occlusion". *Biomed Opt Express* 11: 4560–4570 (2020).

60. F Knieling, C Neufert, A Hartmann, et al. "Multispectral optoacoustic tomography for assessment of Crohn's disease activity". *N Engl J Med* 376: 1292–1294 (2017).

61. LV Wang, J Yao. "A practical guide to photoacoustic tomography in the life sciences". *Nat Methods* 13: 627–638 (2016).

62. EM Strohm, MJ Moore, MC Kolios. "Single cell photoacoustic microscopy: A review". *IEEE J Sel Top Quantum Electron* 22: 137–151 (2016).

63. W Groner, JW Winkelman, AG Harris, et al. "Orthogonal polarization spectral imaging: A new method for study of the microcirculation". *Nat Med* 5(10): 1209–1212 (1999).

64. O Goertz, A Ring, A Kohlinger, et al. "Orthogonal polarization spectral imaging: A tool for assessing burn depth?" *Ann Plast Surg* 64(2): 217–221 (2010).

65. CE Virgini-Magalhaes, CL Porto, FF Fernandes, et al. "Use of microcirculatory parameters to evaluate chronic venous insufficiency". *J Vasc Surg* 43(5): 1037–1044 (2006).

66. AA Kamshilin, VV Zaytsev, AV Lodygin, et al. "Imaging photoplethysmography as an easy-to-use tool for monitoring changes in tissue blood perfusion during abdominal surgery". *Sci Rep* 12: 1143 (2022).

67. G Saiko. "What remote PPG oximetry tells us about pulsatile volume?" *Biomedicines* 12: 1784 (2024).

68. W Verkruysse, M Bartula, E Bresch, et al. "Calibration of contactless pulse oximetry". *Anesth Analg* 124(1): 136–145 (2017).

69. T Burton, G Saiko, A Douplik. "Feasibility study of remote contactless perfusion imaging with consumer-grade mobile camera". *Adv Exp Med Biol* 1395: 289–293 (2022).

70. SP Schraven, B Kossack, D Strüder, et al. "Continuous intraoperative perfusion monitoring of free microvascular anastomosed fasciocutaneous flaps using remote photoplethysmography". *Sci Rep* 13: 1532 (2023).

CHAPTER 6

Scattering-Based Physiological Optical Imaging Techniques

LIGHT SCATTERING REFERS TO a broad range of physical processes where electromagnetic waves deviate from straight trajectories due to localized non-uniformities. Light scattering in biological tissues occurs predominantly on the intracellular and extracellular objects with mismatched refractive indices: nuclei, organelles, collagen fibers, etc.

More particularly, scattering in turbid tissues arises from mismatches in the refractive indices between intracellular or extracellular fluid and the cellular components, which include connective tissue fibers, cytoplasmic organelles, cell nuclei, and melanin granules. These components appear to blend at a microscopic level, with variations in the refractive index, which ranges from 1.35–1.36 (extracellular fluid) and 1.360–1.375 (cytoplasm) to 1.38–1.41 (nucleus, mitochondria, and organelles) to 1.46 (cell membrane) to 1.6–1.7 (melanin) [1].

Based on the energy transfer (or lack thereof) between the photon and the media, the scattering phenomena can be divided into inelastic and elastic scattering, respectively. Elastic scattering is the predominant mechanism of electromagnetic wave scattering in turbid media. In this type of scattering, the energy transfer between the photon and the scattering center does not occur. As a result, the incident and scattered photons have

DOI: 10.1201/9781003482505-8

the same energy. The lifetime, τ_e, of the elastic scattering process is usually very short on the order of 10^{-15} s. In this chapter, we will consider elastic scattering-based techniques. Inelastic (Raman) scattering will be considered in Chapter 9.

The complex nature of scattering in turbid tissues can be understood by examining various biological structures (such as cells and fibers) and levels of their organization (e.g., macrostructures like blood vessels). Tissue structures vary in size from tens of nanometers to hundreds of micrometers. Mammalian cells, for example, are in the range of 5–75 micrometers. The epidermal layer of the skin is composed of large, uniformly sized cells. In contrast, fat cells can vary widely in size, from a few micrometers to 50–75 micrometers in normal cases and 100–200 micrometers in pathological conditions. Within cells, there are a variety of structures that affect how the tissue scatters light, such as cell nuclei (5–10 micrometers in diameter), mitochondria, lysosomes and peroxisomes (1–2 micrometers), ribosomes (20 nanometers) and structures within organelles that can be up to a few hundred nanometers.

As a result of this complex organization, both Rayleigh and Mie scattering occurs in realistic tissues. Typically, Mie scattering is dominant for wavelengths of 500 nm and longer. However, in the range shorter than 500 nm, the contribution of the Rayleigh scattering can be pretty significant.

Even though optical coherence tomography (OCT) is based on light scattering, as an inherently microscopic technique, it will be considered in Chapter 12 (see section 12.2.3).

As such, all the techniques considered in this chapter are dynamic light scattering techniques, where the temporal fluctuations of the recorded signal are analyzed to extract functional information.

6.1 MICROCIRCULATION IN SUPERFICIAL TISSUES

Scattering-based techniques have an established clinical utility for investigating microcirculation in superficial tissues. Two primary technologies, Laser Doppler Imaging, and Laser Speckle Imaging, are noncontact modalities that exploit reflectance geometry.

6.1.1 Laser Doppler

Photons change frequency during scattering on mobile scatterers (the Doppler effect). Red blood cells (RBCs) are primary mobile scatterers in cutaneous circulation; thus, the Doppler effect can be used to evaluate their velocities. As the velocity of RBCs in microcirculation is several mm/s, the effect is small and can be detected only using coherent light.

Laser Doppler Flowmetry (LDF) is a noninvasive technique that utilizes the optical Doppler effect to assess microcirculatory blood perfusion. In practice, LDF often refers to a single-point measurement with a fiber-optic (contact) probe, which is an experimental modality. In contrast, Laser Doppler Imaging (LDI) or Laser Doppler Perfusion Imaging (LDPI) refers to a clinical imaging modality.

In typical LDI/LDF implementation, the tissue is illuminated with a monochromatic light source with a long coherence length (laser). The light undergoes scattering on static and dynamic tissue scatterers. Static and dynamic light fields interfere, and for typical blood cell flow speeds of a few mm/s, they produce a detectable beating of the intensity with a frequency in the kHz range. The response is typically analyzed in the frequency domain. It has been established (see, for example, [2]) that the concentration c of moving red blood cells is proportional to the zero moment, $c \propto M_0/S(0)$, and that the perfusion P is proportional to the first moment, $P \propto M_1/S(0)$.

The LDI approach has clear advantages over LDF, as a) it is a noncontact modality and b) it provides information over a large area as a problem of single point measurements with LDF is that the cutaneous microcirculation exhibits vast spatial and temporal variation [3].

6.1.1.1 Clinical Utility

Laser Doppler Imaging is a relatively mature diagnostic modality used in many clinical situations. It is primarily used to assess burn depth but is also applied in surgery, wound healing, and general vascular diagnostics.

6.1.1.1.1 Burns   Burn injuries can be categorized into four stages according to their depth: superficial epidermal, superficial dermal, deep dermal, and full-thickness burns. It is important to estimate the correct burn depth early after the injury, as it significantly influences the course of treatment. Superficial epidermal and superficial dermal injuries are treated conservatively with dressings. In contrast, deep dermal and full-thickness injuries often require early recognition for prompt surgical excision to accelerate healing and avoid complications. The burn depth assessment is done by visual inspection and is prone to errors even among experienced surgeons [4]. In particular, almost a third of clinical assessments of intermediate burn depth overestimate the extent of the injury, potentially resulting in unnecessary surgery [5], which prompts the development of more objective tools.

Doppler techniques are able to differentiate between deep (low or lack of perfusion) and superficial (higher perfusion) burns. LDI advantages over LDF (noncontact modality, which visualizes a large field of view) are significant for burns. As such, LDI has been tried for burn depth assessments [5]. However, despite LDI being well established for burn depth assessment, it is infrequently employed [6].

6.1.1.1.2 Wound Care   LDI can be used to assess wound healing potential. In particular, sufficient perfusion is an important precursor of wound healing. In addition, it is known that in healing wounds, the perfusion is particularly high at the periphery of the wound due to the inflammatory reaction associated with the normal wound-healing process.

In a comparison study [7], Laser Doppler Flowmetry and $TcpO_2$ were performed in 25 patients with chronic lower limb ulcers with various etiology (experimental arm) and 25 healthy individuals (control arm). A statistically significant difference ($p < 0.05$) was found between the LDF values of the two groups. However, no statistically significant differences were found between the two groups via the $TcpO_2$ measurements.

One study [8] used LDI to measure the circulation in ischemic ulcers. Ten out of the 25 included patients were diabetic foot patients. They found that changes in the circulation measured by LDI may coincide with changes in the number of visible capillaries within an ischemic ulcer.

Some studies used LDI to measure the effect of different healing techniques for diabetic foot ulcers and showed that LDI could measure the circulation in diabetic foot ulcers and that this technique could be used to assess microcirculation [9].

6.1.1.1.3 Reconstructive Surgery   LDI is well-documented for free flap perfusion control [10]. In particular, LDI is a practical tool in flap viability assessment during surgery. For example, perfusion imaging allows the surgeon to verify the reestablishment of blood supply during surgery, which is the earliest possible stage when surgical correction is still possible. Post-surgery, LDF was tried to monitor for complications. However, as mentioned before, LDI has advantages over LDF as a noncontact modality, which images large areas rather than a single point measurement (LDF) that is prone to vast spatial and temporal variations.

As perforator flaps have become one of the primary procedures in microsurgical tissue transfer, individual defects require a customized approach to guarantee the most effective treatment. Consequently, the location of prominent perforators becomes highly important and typically requires invasive angiography or Doppler ultrasound investigation. As such, LDI has been tried for perforator identification. Nischwitz et al. [11] compared thermal, hyperspectral, and Laser Doppler imaging to detect the deep inferior epigastric artery perforators (DIEP flaps). They found the lack of performance by LDI as a positive match for 94.44%, 38.89%, and 0% of patients and 92.59%, 25.93%, and 0% of images for the thermal, hyperspectral, and LDI methods, respectively, compared to the handheld Doppler ultrasound.

6.1.1.2 Laser Doppler: Practical Considerations

Modern laser Doppler imagers use lasers in the near-infrared spectrum (typically, 808 nm wavelength) to increase penetration depth and reduce sensitivity to the skin tone [12].

Earlier LDI tools were implemented as a scanning modality, which assesses blood perfusion over large areas >1,000 cm^2. However, mapping the blood flow with a scanning Laser Doppler imager usually takes minutes, often requiring the subject to be stabilized during imaging. High susceptibility to movement-induced artifacts, combined with the long recording time, poses clinical challenges for the testing of children or individuals with movement disorders such as Parkinson's [6].

However, full-field LDI systems using fast-speed cameras have been developed and adopted in recent years. For example, Leutenegger et al. [13] reported a full-field LDI system capable of monitoring blood perfusion in an area up to 50 cm^2 with 480 × 480 pixels per frame at a rate of 12–14 frames per second. However, this decrease in acquisition time came with a demand for high-speed cameras capable of acquisition at 25k frames per second. Thus, the current trade-off is between slow acquisition using scanning methodology or expensive full-field setups.

It should be pointed out that full-field illumination results in a speckle pattern, which, to detect intensity fluctuations, needs to be resolved in time and less demanding in space [13].

It is known that laser Doppler measurements demonstrate spatial and temporal variations. As such, it is recommended [14] that measurements are taken under the same conditions where appropriate (e.g., same site studied, same ambient temperature, complete acclimatization, caffeine-, nicotine- and vasodilator-free, etc.), with the same LDI parameters (e.g., wavelength, scanning speed, scanning distance, DC values, and image normalization) to ensure valid comparisons.

In addition, temporal variations in blood flow over short periods (days to weeks [15]) and more extended periods (due to seasonal changes [16]) may need to be taken into account, as does the effect of temperature variation, physical and mental activity, and consuming certain vasoactive substances (for example caffeine), which have been shown to have a significant impact on the laser Doppler technique [15].

6.1.2 Laser Speckle Imaging

A speckle pattern is formed by reflecting coherent light from a rough surface or by reflecting or transmitting the light through a medium with refractive index distribution. This phenomenon results from the interference of different reflected beams with random relative optical phases.

If the object does not move and the coherent light source (laser) is stable, the interference pattern does not change over time (a static speckle pattern). If an object (e.g., particle) moves within the illuminated area, the interference pattern will change in time (a dynamic speckle pattern).

As red blood cells (RBCs) are the primary moving scatterers in the body, determinations of RBC velocity may be obtained by assessing the statistical behavior of speckle patterns.

Laser Speckle Imaging refers to a large group of technologies developed since 1972. Initially, the technology was based on traditional photography with 35 mm photographic films. In particular, Archbold and Ennos [17] developed double-exposure speckle photography, which was based on the principle that a photographic record of two correlated and mutually shifted speckle structures gives rise to parallel straight fringes in the Fourier plane (Young's fringes). The spacing and orientation of these fringes depend on the displacement of the object and its direction between both photographs. Despite the fact that Iwai and Shigeta [18] developed a digital version of double-exposure speckle photography, the processing and interpretation of fringes remained quite complicated.

A more direct approach was used by Fercher and Briers [19], who developed single-exposure speckle photography with a long exposure time.

With the advances in digital cameras, laser speckle imaging became digital, and practical solutions for medical applications emerged [20].

6.1.2.1 Clinical Utility

Laser Speckle Perfusion Imaging, LSPI, and Laser Speckle Contrast Imaging, LSCI, have been shown to estimate reliably and visually represent tissue perfusion [21]. As such, LSPI/LSCI is a well-established research modality used to study microcirculation in different types of tissues, including retina [22], superficial tissues, and neurophysiology [23].

For superficial tissues, LSPI/LSCI has been translated to clinical practice. In particular, LSPI/LSCI is a relatively mature diagnostic modality used in many clinical situations, including burns, wound care, and reconstructive surgery. Laser Doppler Imaging and Laser Speckle Imaging can be used interchangeably in all these applications. As the medical context of these indications and the utility of perfusion imaging were discussed in section 6.1.1, here we will refer to certain LSPI/LSCI studies.

6.1.2.1.1 Burns   As a fast, noncontact, large filed modality, LSPI is particularly helpful in assessing burn depth. Animal studies have shown that LSPI/LSCI is able to demonstrate significant blood flow discrepancy among different burn severities,[4] especially emphasized for the challenging differentiation between superficial (dermal) and deep dermal thermal injuries [24].

6.1.2.1.2 Wound Care   LSPI is used for the noninvasive assessment of the blood flow of cutaneous wounds [25]. For example, Hellmann et al. [26] used Laser Speckle Imaging to study whether iontophoresis increases skin microcirculation in diabetic patients (iontophoresis is a process of transdermal drug delivery). Typically, a vasodilator such as acetylcholine was used to measure a tissue's "microvascular capacity." In particular, the authors found that LSCI could be used to diagnose small changes in microcirculation in specific areas, such as the skin on the foot or ankle.

6.1.2.1.3 Reconstructive Surgery   LSCI has been used to test for free flap perfusion control [27, 28]. For example, McGuire et al. [27] used LSCI to study the pattern of ischemia and return of revascularization in three surgical models of wound ischemia, namely a cranial-based myocutaneous flap, an identical flap with underlying silicone sheeting to prevent engraftment, and a complete incisional flap without circulation.

6.1.2.2 Practical Considerations

Current laser speckle imaging approaches analyze the statistical properties of speckle patterns in spatial or temporal domains.

Laser Speckle Contrast Analysis (LASCA) [29] emerged as a practical way to analyze speckle patterns in the spatial domain. This approach analyzes the per-pixel intensity distributions in blocks of NxN pixels (typically, 5x5 or 7x7). The drawback of this technology is lower spatial resolution.

Laser Speckle Imaging (LSI) [30] analyzes speckle patterns in the time domain. This technique calculates the contrast per pixel based on a time sequence. The drawback of this technology is its lower temporal resolution.

A modern version of the technology, often termed Laser Speckle Perfusion Imaging (LSPI) [31], combines the best of two worlds, allowing temporal and spatial analysis. If the time resolution is required, then the speckle pattern is analyzed in the spatial domain. If the spatial resolution is required, then the speckle pattern is analyzed in the time domain.

The sampling depth of laser speckle methods has not yet been well established. While some authors [32] estimate a sampling depth of 300 μm, others [33] suggest that more than 95% of the scattering that produces the speckle pattern occurs in the outermost 700 μm of vascular tissue. However, the sampling depth is wavelength dependent.

Laser Speckle Imaging is often compared to Laser Doppler Perfusion Imaging as both are the primary modalities for tissue perfusion visualization. Laser Speckle Imaging has several advantages over Laser Doppler Perfusion Imaging. Historically, it was a speed of acquisition. As early LDPI technologies were scanning based, they required a long acquisition time (on a scale of minutes). With advances in digital sensors, modern LDPI are implemented as full-view modalities. Thus, this advantage is reduced to the cost of the camera. In particular, to perform LASCA measurements, an inexpensive camera with a 200 Hz frame rate (i.e., an integration time of 5 ms) is sufficient, while for LDPI, a state-of-the-art high-speed camera with a 25 kHz frame rate is needed.

The disadvantage of the Laser Speckle Imaging approach is that, unlike Laser Doppler Imaging, Laser Speckle Imaging lacks consistent models for interpreting physiological information. For example, a model linking the measurement outcome to the perfusion is absent. As such, the concentration of RBCs cannot be extracted from the recorded patterns [20]. Moreover, multiple assumptions need to be made. For example, statistical properties of velocity distribution (Lorentzian, Gaussian, or their combination (Voigt)), the fraction of moving red blood cells, and other parameters (e.g., particle size) need to be explicitly specified.

However, studies [34] suggest that quantitative measurement of blood flow velocities using LSCI is feasible. As such, in everyday clinical situations, speckle imaging proves to be on par with the more established LDI for functional perfusion testing [6]. For example, testing LSCI against LDI reveals a linear correlation of the results after provocation tests [35, 36], excellent interday reproducibility [35], and good correlation in burn depth assessment [37].

6.2 MORE COMPLEX GEOMETRIES

In addition to reflectance mode, which can be used for superficial tissue investigations, several technologies were proposed to investigate tissue properties in other geometries.

6.2.1 Speckle Plethysmography

Ghijsen et al. [38] extended traditional applications of laser speckle imaging and proposed speckle plethysmography (SPG) as a wearable technology for characterizing microvascular flow and resistance. The device, termed Affixed Transmission Speckle Analysis (ATSA) (Flowmet, LAS, Inc., Irvine, CA), uses a 785 nm laser diode placed opposite to a 752-pixel x 480-pixel CMOS camera that captures 200 fps in transmission geometry. It has the form factor of a pulse oximeter and can capture laser speckle signals across fingers. The image processing pipeline, termed speckle plethysmography (SPG), allows extraction of two physiological signals: one related to pulsatile vascular expansion (the photoplethysmographic (PPG) waveform) and one related to pulsatile vascular blood flow, which authors named the speckle plethysmographic (SPG) waveform. The authors extracted the waveforms from 16 volunteers of different ages and observed the age-related dependence of the time delay between PPG and SPG waveforms. Moreover, they analyzed the shape of the SPG waveform using the harmonic content at heart rate frequency (using the ratio of the third harmonic to the first harmonic) and also found age-related dependence. Based on these findings, they associated the time delay and harmonic content changes with arterial stiffness and vascular aging.

Razavi et al. [39] compared SPG with the ankle–brachial index (ABI), toe–brachial index (TBI), and clinical presentation of patients per Rutherford category on 167 limbs (90 patients). Qualitatively, they found that SPG is analogous to Doppler velocity measurements, and waveform phasicity and amplitude degradation were observed with increasing PAD severity.

6.2.2 Diffuse Correlation Spectroscopy

Unlike traditional dynamic light scattering approaches, which consider mostly single scattering events, Boas et al. [40] developed its multiple scattering analog. Their approach, called "diffuse correlation spectroscopy" (DCS), is based on temporal correlations. In particular, the authors found that the transport of temporal correlation through heterogeneous turbid media can be viewed as a scattering of diffuse correlation.

The authors [40] also proposed a 3D (tomographic) image reconstruction approach, which can be used to visualize heterogeneities with different scattering properties. In particular, they were able to visualize a 1.3 cm diameter spherical cavity filled with an aqueous suspension of 0.296 μm polystyrene balls in a homogeneous solid cylinder (collinear with cavity) of TiO_2 suspended in resin with a diameter of 4.6 cm. Six hundred measurements were made every 30° at the surface of the cylinder for g = 0, 1, and 2 cm, with source-detector angular separations of 30° and 170° and correlation times of τ= 15, 25, 35, 45, 55, 65, 75, and 85 μs.

6.3 CONCLUSIONS

Light scattering in biological tissues occurs predominantly on the intracellular and extracellular objects with mismatched refractive indices: nuclei, organelles, collagen fibers, etc.

All techniques discussed in this chapter are dynamic light scattering techniques, where the temporal fluctuations of the recorded signal are analyzed to extract functional information.

Laser Doppler Imaging and Laser Speckle Imaging are primary scattering-based techniques translated to medical practice. They are direct methods to measure perfusion and are used to visualize blood flow in superficial tissues. Laser Doppler Imaging is the more established of these two technologies, as it has solid scientific foundations. However, with an accumulation of evidence and comparison with LDI, speckle-based techniques are being actively adopted in research and clinical practices as they are more affordable and straightforward to use and interpret results.

In addition to LDI and LSI, which are used in reflectance geometry on superficial tissues, several other approaches have been proposed to investigate more complex geometries.

REFERENCES

1. VV Tuchin. *Tissue Optics: Light Scattering Methods and Instruments for Medical Diagnosis*. Bellingham, WA: SPIE Press (2007).
2. R Bonner, R Nossal. "Model for laser Doppler measurements of blood flow in tissue". *Appl Opt* 20: 2097–2107 (1981).
3. T Tenland, EG Salerud, GE Nilsson, et al. "Spatial and temporal variations in human skin blood flow". *Int J Microcirc Clin Exp* 2: 8190 (1983).
4. DM Burmeister, A Ponticorvo, B Yang, et al. "Utility of spatial frequency domain imaging (SFDI) and laser speckle imaging (LSI) to non-invasively diagnose burn depth in a porcine model". *Burns* 41: 12421252 (2015).
5. DCG Sainsbury. "Critical evaluation of the clinimetrics of laser Doppler imaging in burn assessment". *J Wound Care* 17: 193200 (2008).
6. JM Rüwald, C Jacobs, S Scheidt, et al. "Laser-based techniques for microcirculatory assessment in orthopedics and trauma surgery: Past, present, and future". *Ann Surg* 270(6): 1041–1048 (2019).
7. E Raposio, N Bertozzi, R Moretti, et al. "Laser Doppler flowmetry and transcutaneous oximetry in chronic skin ulcers: A comparative evaluation". *Wounds* 29(7): 190–195 (2017).
8. ME Gschwandtner, E Ambrózy, B Schneider, et al. "Laser doppler imaging and capillary microscopy in ischemic ulcers". *Atherosclerosis* 3(142): 225–232 (1999).
9. N Morimoto, N Kakudo, PV Notodihardjo, et al. "Comparison of neovascularization in dermal substitutes seeded with autologous fibroblasts or impregnated with bFGF applied to diabetic foot ulcers using laser Doppler imaging". *J Artif Organs* 16(17): 352–357 (2014).
10. S Schlosser, R Wirth, JA Plock, et al. "Application of a new laser Doppler imaging system in planning and monitoring of surgical flaps". *J Biomed Opt* 15: 036023 (2010).
11. SP Nischwitz, H Luze, M Schellnegger, et al. "Thermal, hyperspectral, and laser Doppler imaging: Non-invasive tools for detection of the deep inferior epigastric artery perforators-a prospective comparison study". *J Pers Med* 11(10): 1005 (2021).
12. NC Abbot, WR Ferrell, JC Lockhart, et al. "Laser Doppler perfusion imaging of skin blood flow using red and near-infrared sources". *J Invest Dermatol* 107: 882886 (1996).
13. M Leutenegger, E Martin-Williams, P Harbi, et al. "Real-time full field laser Doppler imaging". *Biomed Opt Express* 2(6): 1470–1477 (2011).
14. AK Murray, AL Herrick, TA King. "Laser Doppler imaging: A developing technique for application in the rheumatic diseases". *Rheumatology* 43(10): 1210–1218 (2004).
15. A Bircher, EM de Boer, T Agner, et al. "Guidelines for measurement of cutaneous blood flow by laser Doppler flowmetry. A report from the standardization group of the European society of contact dermatitis". *Contact Dermatitis* 30: 65–72 (1994).
16. JM Gardner-Medwin, IA Macdonald, JY Taylor, et al. "Seasonal differences in finger skin temperature and microvascular blood flow in healthy men

and women are exaggerated in women with primary Raynaud's phenomenon". *Br J Clin Pharmacol* 52: 17–23 (2001).

17. E Archbold, AE Ennos. "Displacement measurement from double-exposure laser photographs". *Opt Acta* 19: 253–271 (1972).

18. T Iwai, K Shigeta. "Experimental study on the spatial correlation properties of speckled speckles using digital speckle photography". *Jpn J Appl Phys* 29: 1099–1102 (1990).

19. AF Fercher, JD Briers. "Flow visualization by means of single-exposure speckle photography". *Opt Commun* 37: 326–330 (1981).

20. M Draijer, E Hondebrink, T van Leeuwen, W Steenbergen. "Review of laser speckle contrast techniques for visualizing tissue perfusion". *Lasers Med Sci* 24(4): 639–651 (2009).

21. G Mahé, P Rousseau, S Durand, et al. "Laser speckle contrast imaging accurately measures blood flow over moving skin surfaces". *Microvasc Res* 81: 183188 (2011).

22. M Nagahara, Y Tamaki, A Tomidokoro, et al. "In vivo measurement of blood velocity in human major retinal vessels using the laser speckle method". *Invest Ophthalmol Vis Sci* 52: 8792 (2011).

23. P Li, S Ni, L Zhang, et al. "Imaging cerebral blood flow through the intact rat skull with temporal laser speckle imaging". *Opt Lett* 31: 18241826 (2006).

24. C Crouzet, JQ Nguyen, A Ponticorvo, et al. "Acute discrimination between superficial-partial and deep-partial thickness burns in a preclinical model with laser speckle imaging". *Burns* 41: 10581063 (2015).

25. TMAJ van Vuuren, C van Zandvoort, S Doganci, et al. "Prediction of venous wound healing with laser speckle imaging". *Phlebology* 32: 658–664 (2017).

26. M Hellmann, M Roustit, F Gaillard-Bigot, JL Cracowski. "Cutaneous iontophoresis of treprostinil, a prostacyclin analog, increases microvascular blood flux in diabetic malleolus area". *Eur J Pharmacol* 758: 123–128 (2015).

27. PG McGuire, TR Howdieshell. "The importance of engraftment in flap revascularization: Confirmation by laser speckle perfusion imaging". *J Surg Res* 164: e201e212 (2010).

28. T Furuta, M Sone, Y Fujimoto, et al. "Free flap blood flow evaluated using two-dimensional laser speckle flowgraphy". *Int J Otolaryngol* 2011: 297251 (2011).

29. JD Briers, S Webster. "Quasi real-time digital version of single-exposure speckle photography for full-field monitoring of velocity or flow fields". *Opt Commun* 116: 36–42 (1995).

30. H Cheng, Q Luo, S Zeng, et al. "Modified laser speckle imaging method with improved spatial resolution". *J Biomed Opt* 8: 559–564 (2003).

31. KR Forrester, J Tulip, C Leonard, et al. "A laser speckle imaging technique for measuring tissue perfusion". *IEEE Trans Biomed Eng* 51: 2074–2084 (2004).

32. J O'Doherty, P McNamara, NT Clancy, et al. "Comparison of instruments for investigation of microcirculatory blood flow and red blood cell concentration". *J Biomed Opt* 14: 034025 (2009).

33. MA Davis, KSM Shams, AK Dunn. "Imaging depth and multiple scattering in laser speckle contrast imaging". *J Biomed Opt* 19: 086001 (2014).
34. A Nadort, K Kalkman, TG Leeuwen, et al. "Quantitative blood flow velocity imaging using laser speckle flowmetry". *Sci Rep* 6: 25258 (2016).
35. M Roustit, C Millet, S Blaise, et al. "Excellent reproducibility of laser speckle contrast imaging to assess skin microvascular reactivity". *Microvasc Res* 80: 505511 (2010).
36. C Millet, M Roustit, S Blaise, et al. "Comparison between laser speckle contrast imaging and laser Doppler imaging to assess skin blood flow in humans". *Microvasc Res* 82: 147151 (2011).
37. CJ Stewart, R Frank, KR Forrester, et al. "A comparison of two laser-based methods for determination of burn scar perfusion: Laser Doppler versus laser speckle imaging". *Burns* 31: 744752 (2005).
38. M Ghijsen, TB Rice, B Yang, et al. "Wearable speckle plethysmography (SPG) for characterizing microvascular flow and resistance". *Biomed Opt Express* 9: 3937–3952 (2018).
39. MK Razavi, DPT Flanigan, SM White, TB Rice. "A real-time blood flow measurement device for patients with peripheral artery disease". *J Vasc Interven Rad* 32(3): 453–458 (2021).
40. DA Boas, LE Campbell, AG Yodh. "Scattering and imaging with diffusing temporal field correlations". *Phys Rev Lett* 75(9): 1855–1858 (1995).

CHAPTER 7

Fluorescence-Based Physiological Optical Imaging Techniques

After a photon has been absorbed and given its energy to the absorption center (e.g., molecule), there are multiple ways for the absorption center to return to its lowest energy state (ground state), the most common at room temperature. Three common pathways are: a) to induce radiation emission such as fluorescence, b) to initiate energy confinement leading to temperature rise, and c) to drive a chemical reaction (intra- and intermolecular energy transfer, isomerization, dissociation, and ionization).

If it results in photon emission, the process is known as luminescence. Fluorescence and phosphorescence processes are two primary types of luminescence. They are illustrated in the Jablonski diagram (see Figure 7.1).

When a photon is emitted from an excited state to a lower energy state with the same spin state, for example, $S_1 \rightarrow S_0$, the process is referred to as fluorescence. On the other hand, if the emission occurs between different spin states, for example, $T_1 \rightarrow S_0$, it is called phosphorescence. Fluorescence is much more likely to occur than phosphorescence, resulting in shorter lifetimes for fluorescent states (ranging from 10^{-6} to 10^{-10} seconds) compared to longer lifetimes for phosphorescent states (ranging from 10^{-4} seconds to minutes or even hours). Additionally, non-radiative processes can occur, including internal conversion, intersystem crossing, and vibrational

DOI: 10.1201/9781003482505-9

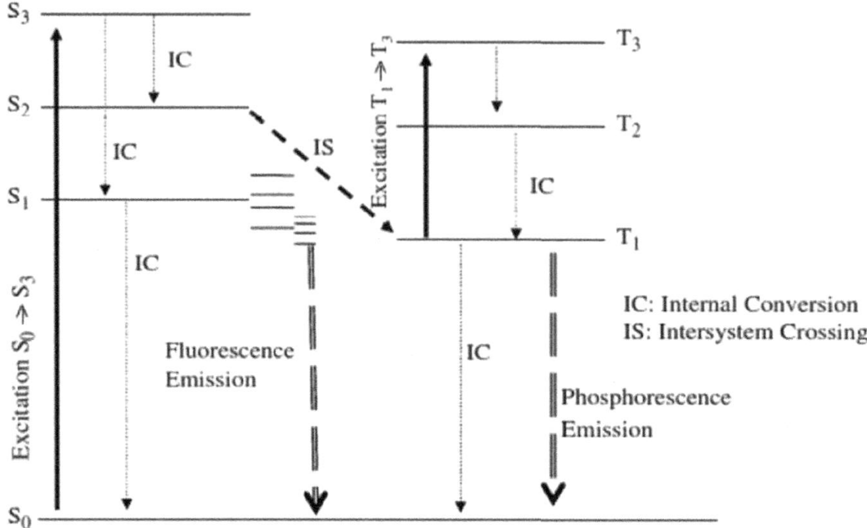

FIGURE 7.1 Jablonski diagram. Absorbed energy from S_3 can return to the ground state via internal conversions (IC, with or without light emission) or intersystem crossing (IS) to the triplet state.

relaxation. These processes can be seen in Figure 7.1, where internal conversion is the transition between energy states of the same spin without the emission of radiation. Intersystem crossing is the transition between different spin states without the emission of radiation.

Fluorescence is a property of emitting light of a longer wavelength on the absorption of light energy, which essentially co-occurs with the excitation of a sample. Unlike inelastic scattering (see Chapter 9), the fluorescence output strongly depends on the energy (wavelength) of the incident light. As such, fluorescence is characterized by both the emission and excitation spectra. In particular, an emission spectrum represents the light emitted by a luminescent material at various wavelengths when a specific, limited range of shorter wavelengths stimulates it. The excitation spectrum is the emission spectrum monitored at one wavelength, with the intensity at that wavelength being measured in relation to the wavelength used to excite it.

Fluorescence effectiveness is characterized by quantum yield, QY, which is defined as the proportion of photons emitted by fluorescence compared to the photons absorbed.

Brightness is another important metric which characterizes both absorption and emission of the fluorophore. It is defined as a product of the molar extinction coefficient [$cm^{-1}M^{-1}$] and quantum yield.

Fluorophores can be broadly classified into three major types: inorganic molecules (pigments), organic molecules (dyes), and nanostructures/nanoparticles.

Autofluorescence (AF) is the natural fluorescence of a material (a tissue) due to the excitation of the endogenous fluorophores in contrast to the fluorescence of a stained material (a tissue or a cell) when exogenous fluorophores are excited.

Native fluorophores in human skin include collagen, elastin, amino acids (tryptophan and tyrosine), NADH, NAD, FAD, and porphyrins. The epidermis contains more tyrosine and tryptophan than the whole skin, so it has a high autofluorescence in the UVA range. Similarly, the psoriatic stratum corneum has a much higher fluorescence than the normal one.

In addition to native fluorescence, some bacteria species produce fluorophores, indicating bacteria presence. In particular, *P. aeruginosa* produces siderophore pyoverdin, which fluoresces in the blue-green range (peak at 465 nm), while more than 20 other species (including clinically relevant *S. aureus*) produce porphyrins, which fluoresce in the orange-red range (600–660 nm).

Unlike endogenous fluorescence techniques based on tissue autofluorescence, exogenous fluorescence techniques use external agents to incite fluorescence. Several external fluorescent agents, like fluorescein or indocyanine green (ICG), are used to visualize perfusion in tissues using exogenous fluorescence.

We will focus only on imaging modalities as in other parts of the book. However, it should be mentioned that fluorescence-based techniques are used in many other clinical modalities. For example, Profusa Inc (TX, USA) developed a method for tracking tissue oxygen in real time with injectable, tissue-integrating microsensors (phosphorescent long-term oxygen sensors) [1]. The system consists of a small (500 mm × 500 mm × 5 mm), soft, flexible, tissue-like sensor from biocompatible hydrogels and a handheld or wearable Bluetooth optical reader for intermittent or continuous noninvasive monitoring. The sensors are based on oxygen-dependent quenching of palladium porphyrin phosphorescence and are engineered to function for months to years in the body. In particular, dyes are synthetically engineered to emit in the near infra-red (optical window of the skin), covalently linked to the hydrogel backbone, and be stable under *in vivo* conditions. The system was able to monitor tissue oxygen for nine months in a Sinclair mini-pig model [2]. The method can be handy for the early detection of restenosis after revascularization procedures.

Also, this chapter will discuss traditional (single photon) fluorescence. Two-photon (or multiphoton in general) fluorescence will be considered in Chapter 12 (section 12.2.1.2). The reason for that is twofold. Firstly, two-photon microscopy (TPM) encompasses two techniques, which often are used in conjunction with one another: two-photon excitation fluorescence (TPEF) and second harmonic generation (SHG), which, technically speaking, is not a fluorescence technique. Secondly, as the name implies, TPM is a microscopic technique similar to confocal microscopy.

7.1 EXOGENOUS FLUORESCENCE IMAGING

Exogenous fluorescence imaging refers to fluorescence-based technologies, which use an external fluorescent agent. Unlike endogenous fluorescence, bioprobes can be tailored to meet specific requirements. The general requirements for the exogenous fluorescent agents can be formulated as follows [3]:

- "Photophysics: absorb/emit at wavelengths which pass though the tissue/sample concerned and be differentiated from autofluorescence by Stokes shift or luminescence lifetime.

- Solubility and stability: remain intact and in solution in physiological media for time sufficient for imaging.

- Toxicity: be non-toxic for *in vivo* work and non-cytotoxic for *in vitro* cellular work.

- Uptake and localization: cross lipophilic cell membranes and barriers and localize in the site of interest.

- Sensing action: ideally a probe shows a detectable response to environment reporting on e.g. pH or [O_2] allowing combined imaging and sensing."

Broadly speaking, an exogenous agent can be a dye, inorganic complex, or nanoparticle. We will consider these technologies separately: organic vs. inorganic.

7.1.1 Dye-Based Techniques

Multiple organic dyes (for example, green and red fluorescent proteins, GFP and RFP, respectively) are available for biomedical research. Some of these dyes are cleared for medical use. Among them are fluorescein (498 nm absorption), evans blue (600 nm absorption), methylene blue (660 nm absorption), and indocyanine green (ICG, 800 nm absorption). These dyes are translated into clinical practice in several medical fields.

In the fluorescent angiography procedure, the dye is injected into the bloodstream. Then, the target area is illuminated with the excitation light, and the imaging sensor detects the emitted light. Initially, the method was developed for ophthalmology. Fluorescent angiography (FA) or intravenous fluorescein angiography (IVFA) is a method that involves a specialized camera and a fluorescent dye to investigate the circulation of the retina and choroid. Recently, the technique has been extended to other blood vessels and body parts.

Fluorescence angiography (FA) or indocyanine green angiography (ICGA) is a novel optical modality that visualizes blood flow in vessels and perfusion. It uses an injectable dye, indocyanine green (ICG), injected into the systemic circulation and cleared exclusively via the liver.

In addition to synthetic dyes, several naturally occurring fluorophores are used in clinical practice. For example, hypericin can be used in photodynamic therapy (PDT) and photodynamic diagnostics (PDD) [4].

7.1.1.1 Clinical Utility

Exogeneous fluorescence imaging is used in diverse applications to determine cardiac output, hepatic function, liver blood flow, and ophthalmic angiography. However, its primary utility across different medical fields is perfusion assessment. For example, indocyanine green fluorescence angiography (ICG-FA) is used for real-time intraoperative assessment of anastomotic perfusion to assist intraoperative decision-making and decrease the risk of anastomotic leak. Nguyen et al. [5] investigated the use of ICG-FA for the intraoperative assessment of anastomotic perfusion during bowel resection in oncologic gynecology. They found that ICG-FA enables objective and accurate intraoperative evaluation of anastomotic perfusion in surgeries for gynecologic malignancies.

7.1.1.1.1 Wound Care   ICGA is particularly helpful for patients with peripheral arterial disease (PAD). In [6], the authors analyzed PAD patients

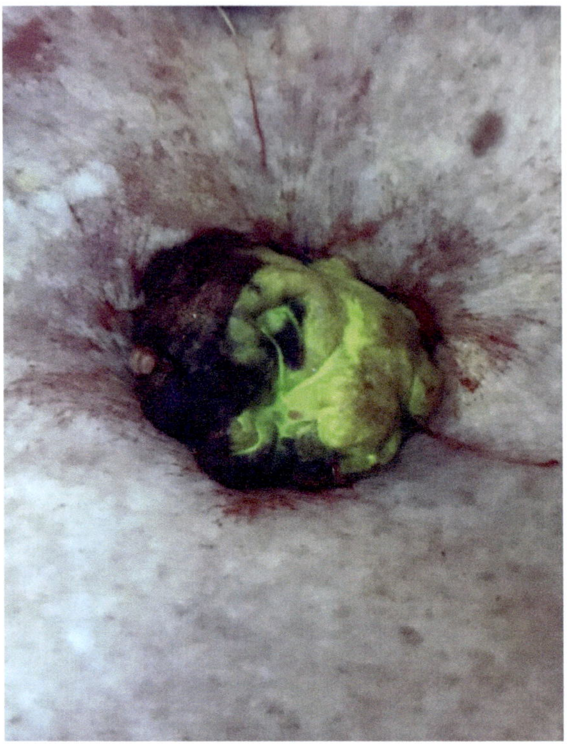

FIGURE 7.2 Indocyanine green imaging of a pressure wound. Indocyanine green was administered via IV to assess the perfusion in a pressure injury wound bed. The image shows that non-perfused areas appear black, while those with adequate perfusion exhibit an intense fluorescence signal. (Photograph courtesy of Dr. Mario A. Martinez Jimenez).

with isolated infrapopliteal lesions who underwent ICGA. They compared the findings with the ankle–brachial index (ABI) and toe–brachial index (TBI) in 14 PAD patients and 9 control patients. The authors found that the Td 90% (the time elapsed from the maximum intensity to 90% of the maximum intensity) correlated most significantly with the ABI value. Thus, the authors concluded that ICGA might be helpful in quantitatively assessing peripheral perfusion.

7.1.1.1.2 Oncology Fluorescence imaging (FI) for cancer therapy and cancer gene therapy was reviewed by Woo et al. [7]. In addition to perfusion assessment, FI is used in oncology for sentinel lymph node detection, intraoperative tumor delineation, and lymphatic obstruction mapping.

ICG is the most commonly used dye in oncologic FI applications. Its utility is based on the accumulation of ICG in tumors. However, ICG does not target cancer specifically. Instead, ICG exhibits passive extravasation through loose blood vessels around cancer cells and accumulates within cancer cells. This phenomenon is called the enhanced permeability and retention effect [8].

Many cancers, including breast, colon, and skin cancers, spread to lymph nodes. Such lymphatic spread has significant prognostic and therapeutic implications, as those patients with lymphatic metastases have poorer survival and usually benefit from excision of the affected lymphatic tissues [5]. Fluorescence imaging is one of the methods to identify sentinel lymph nodes (SLN) [9].

Fluorescent imaging is also beneficial for the mapping of lymphatics. In particular, many patients after cancer surgery develop lymphedema, which is swelling of extremities near areas of sites of surgery caused by the destruction of lymphatics. Lymphatic bypasses emerged as an essential therapeutic procedure to reduce morbidity and improve the quality of patients' lives. ICG-based lymphatic mapping is invaluable in the surgical execution of these bypasses, improving outcomes [10].

In addition, specific fluorescent agents for cancer localization are being developed. Generally, they fall into two categories [6]: antibodies or ligands for proteins and receptors on cancer cell surfaces and substrates for cancer-specific metabolic pathways. Typically, they are labeled with either ICG or one of the cyanine red dyes, including Cy5, Cy5.5, and Cy 7.0.

Fluorescence imaging was also applied to oncolytic viral therapies, cancer staging, and gene therapy tracking. These approaches rely on viruses carrying transgenes coding GFP. In particular, oncolytic viral therapy aims to design viruses that specifically infect and kill cancer and generate an immune response directed at cancer. By designing viruses that carry a transgene coding a fluorescent protein, fluorescence imaging allows documentation of sites of infection and replication.

7.1.1.2 Practical Considerations

Organic dyes are highly emissive, which enables single-molecule detection [11]. Resolution down to 15–20 nm can be obtained in the focal plane using special techniques such as stimulated emission depletion microscopy [12]. Organic dyes typically have singlet-state transitions S_1->S_0 (see Figure 7.1); thus, they are characterized by very short fluorescence lifetimes (nanoseconds and less). The downside of organic dyes is photobleaching.

While most fluorescence imaging techniques started from fluorescein, nowadays, most techniques are ICG-based. ICG has a documented record of safe clinical use. It allows high SNR, as it emits in the NIR spectrum range (820 nm), where the tissue autofluorescence is reduced. ICG is known to have a skin penetration depth of approximately 1 cm. Another important feature of ICG is that it does not leak into extravascular space, allowing for multiple images in the same patient settings [13].

7.1.2 Inorganic and Nanoparticle-Based Techniques

In this section, we will consider inorganic and nanoparticle-based techniques.

The physics of nanoparticles is very different from the physics of bulk materials. For example, in semiconductors, when the size of the nanoparticles becomes smaller than a critical value known as the exciton Bohr radius (typically 10 nm), 3D confinement of charge carriers occurs, limiting the number of energy states in the valence and conduction bands and resulting in optical properties that can be tuned by varying the particle size or the internal chemical composition.

There are several groups of nanoparticle-based approaches.

7.1.2.1 Quantum Dots

New materials, like quantum dots (QDs), represent a promising direction in nanomedicine. QDs, also called semiconductor nanocrystals, are semiconductor particles 2–10 nm in size, having optical and electronic properties that differ from those of larger particles due to quantum mechanical effects. They can be considered 0D objects confined from all three spatial directions. When the quantum dots are illuminated by UV or visible light, an electron in the quantum dot can be excited to a state of higher energy from the valence band to the conductance band.

QDs have unique properties that offer great utility to fluorescence imaging.[14] (1) Specificity – the synthesis of QDs results in organic capping ligands that make them biocompatible and suitable for biological targeting development. This is achieved by surface modification and linking with antibodies, peptides, or small molecules. The QDs can be modified to become specific ligands that discretely couple with their target. (2) Adjustable emission – narrow emission band and size-tunable Gaussian emission spectra in the visible and NIR spectra. (3) Strong signals – QDs have a high fluorescence quantum yield and a large Stoke's shift.

However, QDs have several drawbacks: 1) flickering of the emission when there are only a small number of QDs in the target material, and 2) concern about their long-term toxicity, for instance, through the release of toxic cadmium, arsenide, or selenide ions [15].

7.1.2.2 Carbon Nanostructures

Multiple carbon-based nanostructures have been proposed since the discovery of carbon dots (CDs) in 2004 [16], including carbon nanotubes (CNTs), graphene and its derivatives, fullerene, nanodiamond, and graphene quantum dots (GQDs).

The carbon nanoparticles have several favorable characteristics, including low cost for synthesis/fabrication, high photostability, tunable emission, good intracellular solubility, chemical inertness, and nontoxicity.

Carbon nanostructures emit light from blue to NIR range of spectra. The optical properties of carbon nanostructures can be modified by varying their size. They also depend on the synthesis procedure and the kind of surface passivation.

Carbon dots (CDs) can be modified either in the core structure by partial substitution of carbon with other elements (i.e., doping by nitrogen, sulfur, or boron) or by surface functionalization [17]. Functionalization of CD's surface is essential to 1) enable target-specific imaging and 2) enhance their fluorescence QY, as they have lower QY than semiconductor quantum dots.

7.1.2.3 Transition Metal and Lanthanide Complexes and Nanoparticles

Many d-block luminophores have been used in cell imaging, mainly d^6 complexes (Re^I, Ru^{II}, Ir^{III}), Pt^{II}, and Au^I compounds [3]. Broad luminescence bands mainly arise from metal-to-ligand charge-transfer (^{3}MLCT) or metal-to-metal charge-transfer (^{3}MMCT) states [18].

Both long luminescence lifetimes and large Stokes shifts are associated with triplet excited states (see Figure 7.1) caused by spin-orbit coupling effects in complexes of the heavier elements. As a result, the second and third transition series and f-electron-based compounds (lanthanides) dominate in this area.

The main current applications of lanthanides in biology and medicine are in magnetic resonance imaging (MRI), as contrast agents (almost exclusively Gd^{III} complexes), and luminescent assays (immunoassays, protein staining). However, f-electron-based compounds (lanthanides) become commonly used in fluorescent imaging as they have several advantages over other fluorophores. In particular, they have tunable emission in a broad spectral range (from UV to NIR) and a low propensity for photobleaching.

The unique emission properties of lanthanide ions arise from intra-configurational 4f–4f transitions, which are orbitally forbidden. It has two major implications. Firstly, the excited state lifetime is long enough (in the microsecond–millisecond range) to allow time-resolved techniques. Secondly, the absorption coefficient of lanthanide ions is very low. The common approach to address it is to enhance absorption by so-called "luminescence sensitization," where an antenna or a chromophore is grafted to the surroundings of the lanthanide species. The antenna efficiently absorbs UV-visible radiation and transfers it to the emissive excited state of the lanthanide ion.

Due to their unique properties, lanthanide complexes enable time-resolved detection (TRD) and Förster resonance energy transfer (FRET). Additionally, unlike many other fluorophores and luminophores, Ln^{III} emission bands are almost unaffected by the chemical environment [19].

As it is challenging to conjugate inorganic molecules with biomolecules, the bioprobe development for transition metal and lanthanide complexes can be simplified by using nanoparticles. Multiple nanoparticles were developed for bioimaging purposes. They can be divided into downshifting and up-converting nanoparticles (UCNP). Downshifting refers to the absorption process of a high-energy photon, followed by the emission of a lower-energy one (Stokes'emission). In up-conversion (anti-Stokes fluorescence), two or more long wavelength photons excite the nanostructure, emitting a shorter-wavelength photon. This process should not be confused with the two-photon process, a separate nonlinear physical phenomenon.

7.1.2.4 Clinical Utility

The metal complexes and nanoparticles are widely used as fluorescent agents in immunoassays. Thus, fluorescence imaging technologies can be helpful in the automation of high-throughput immunoassays in clinical laboratory settings.

Nanoparticles' *in vivo* and *ex vivo* use is based on their important property to accumulate in diseased tissues, e.g., tumors. In addition, the growing trend in fluorescent imaging is cancer cell targeting using targeting moieties on the nanostructure surface. Such targeting moieties can provide the internalization of nanoparticles into cancer cells and tissues through a ligand-receptor interaction. Highly specific cancer cell targeting and efficient distribution avoid the side effects of nonspecific bonding. Various ligands (peptides, antibodies, aptamers, or small molecules) and targeting strategies have been proposed to improve tumor targeting and penetration of nanomaterials, directing them to tumors.

Bioprobes can be tailored to sensing microenvironment, for example, pH or $[O_2]$ (see, for example, [3]). In addition, they can be designed to detect halides, nitrate, phosphates (including HPO_4^{2-}, HPO_4^{-}, adenosine mono-, di-, and triphosphates), acetate, oxalate, malonate, succinate, lactate, citrate, urate, amino acids [20]. They can easily integrate multi-analyte detection.

7.1.2.5 Practical Considerations

Inorganic probes offer some very distinct advantages over organic dyes. For example, they are characterized by small, even nonexistent photobleaching. As the organic dyes mostly have emission in the visible spectrum range, this range is suboptimal, as it has strong absorption and scattering within the tissue and strong tissue autofluorescence. Nanoparticle-based approaches may overcome these issues by shifting the excitation and emission in the NIR range. It can be achieved by three methods: 1) two-photon nonlinear processes, 2) upconversion, and 3) designing nanoparticles with NIR excitation and emission. As the two-photon nonlinear method requires high-power light, it is limited to *ex vivo* and *in vitro* applications, as "tissue-friendly" irradiation should remain below 300 mW/cm^2 to avoid heating damages.

Another advantage of nanoparticles (particularly transition metal and lanthanide-based) is their longer fluorescence lifetime, which enables time-resolved detection (TRD) that considerably enhances signal-to-noise ratio by eclipsing autofluorescence from endogenous biomolecules.

Persistent luminescence nanoparticles are an interesting possibility as well. These are typically transition metal or lanthanide-based particles whose luminescence may stay for a long time (e.g., hours). This type of particle allows for the excitation of particles before injection into the body.

A common disadvantage of all nanoparticles is enhanced non-radiative deactivation due to surface traps and defects that increase with decreasing size, translating into very low quantum yields. However, innovative design, such as adding a passivation shell (core-shell NPs) or doping other non-luminescent metal ions, dramatically enhances the quantum yields, reaching values close to the bulk materials [21]. Another common disadvantage of nanoparticles, which limits their clinical use, is cytotoxicity. Golden nanoparticles and carbon-based nanostructures are an exception here, as these particles are considered to be mostly biocompatible and nontoxic.

As such, we can conclude that in the near term, nanoparticles will likely find much of their role in specialist applications requiring particular properties, such as large Stokes shifts or long lifetimes, rather than replacing organic dyes as the standard fluorochromes [3].

7.2 ENDOGENOUS FLUORESCENCE

While tissue autofluorescence is an undesirable artifact for exogenous fluorescence imaging, it may bring important information about physiology or pathophysiology.

7.2.1 Bacterial Imaging

Bacterial fluorescence imaging is an endogenous fluorescence imaging technique based on the autofluorescence of bacterial metabolites. In particular, most clinically relevant bacteria produce porphyrins [22] (*S. aureus*, *S. epidermidis*, *Candida*, *S. marcescens*, *Viridians streptococci*, *Corynebacterium diphtheriae*, *S. pyogenes*, *Enterobacter*, and *Enterococcus*) or pyoverdine [23] (*P. aeruginosa*), which fluoresce (red and bluish-green fluorescence, respectively) when excited by incident light with a wavelength of 405 nm.

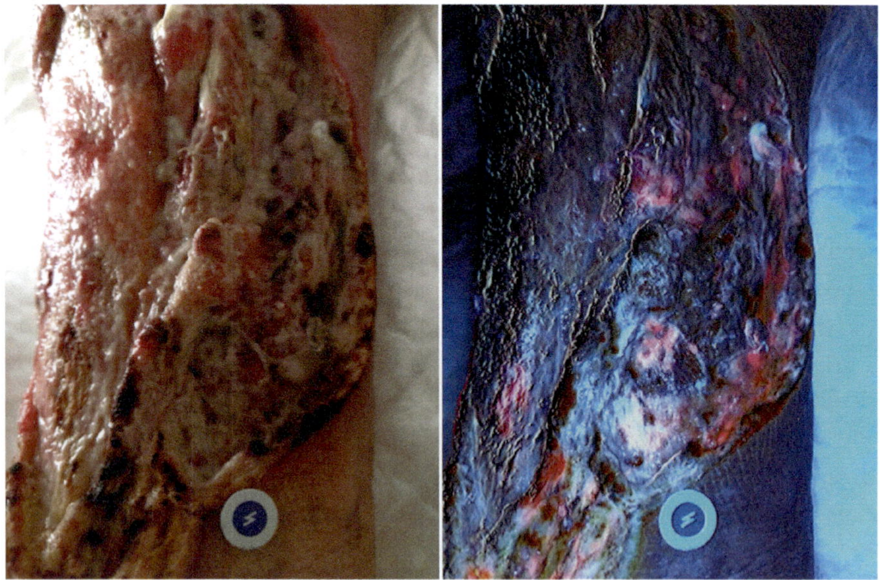

FIGURE 7.3   Bacterial autofluorescence. Bacterial pigments are easily visualized in wound beds after excitation with a 405 nm light source. Species producing heme-containing compounds fluoresce in red, and those producing pyoverdine in cyan. This photograph shows polybacterial contamination, as both cyan and red fluorescence are captured (Photograph courtesy of Jose Ramirez GarciaLuna).

7.2.1.1 Clinical Utility

Bacterial and fungal autofluorescence is used in dermatology and wound care. For example, the Wood lamp is used in dermatology to visualize fungal infections.

Bacterial fluorescence is also translated into the medical practice. As a noninvasive and labor-nonintensive technique, bacterial fluorescence imaging is being adopted in wound care, including point-of-care settings. The utility of fluorescence imaging for visualization of bacterial load in wounds has been demonstrated in numerous clinical studies, including studies with DFUs *in vivo* [24, 25, 26, 27] and *in vitro* [28].

Fluorescence imaging may be used to distinguish between particular strains in the wound (e.g., *P. aeruginosa*), to assess (qualitatively or semi-quantitatively) bacteria present in the wound (infection detection), or to guide sampling, debridement, antimicrobial selection, grafting, and the use of cell tissue products (CTP).

7.2.1.1.1 Infection Detection   The fluorescence imaging demonstrated its utility in identifying bacterial loads of more than 10^4 CFU/g[1] close to real time. In particular, Farhan et al. [29] identified 11 clinical studies (including 613 wounds with various etiology), which aimed to assess the diagnostic accuracy for detecting bacterial loads of $\geq 10^4$ CFU/g and calculated weighted averages for sensitivity (74%), specificity (88%), positive predictive value (91%), negative predictive value (53%), and overall diagnostic accuracy (75%). It represents a three- to fourfold increase in sensitivity compared to clinical signs and symptoms of infection (CSS) alone.

7.2.1.1.2 Sampling and Debridement Guidance   Another practical application of fluorescent imaging is to guide debridement and sampling (swabbing or biopsy). A pilot evaluation compared standard Levine swab results with fluorescence-guided curettage samples and found that the Levine technique gave a 36% false-negative laboratory report [30].

7.2.1.1.3 Treatment Selection   The ultimate utility of fluorescence imaging is to assist in treatment selection. Studies have shown that fluorescence imaging can prompt changes in proposed treatment plans, including alterations in antimicrobial prescribing [31], decisions around negative pressure wound therapy [32], and timing of grafting or applications of cell tissue products (CTP) [33]. Bacterial fluorescence imaging leads to changes in the treatment plan in as many as 73% of cases.

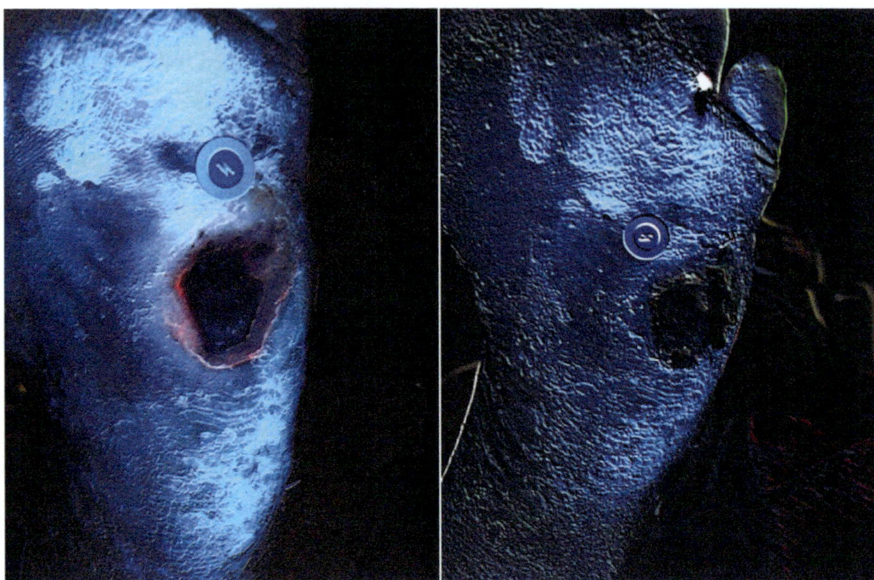

FIGURE 7.4 Targeted debridement using bacterial fluorescence. Bacterial fluorescence can be used to identify areas where excessive bacterial overgrowth is present, as the lower limit of detection of this technology is 10^4 CFU / g organisms. In these images, the left panel shows a wound pre-debridement, and the right one shows the same wound post-debridement. Bright areas of red fluorescence indicate bacteria closer to the wound's surface, while pink areas represent subsurface bacteria. Noteworthy is that keratin exhibits autofluorescence in the white-cyan range. Care must be taken so this signal is not interpreted as bacterial contamination due to pseudomonas. (Photograph courtesy of Jose Ramirez GarciaLuna).

A randomized controlled trial (RCT) compared healing rates and the decision-making associated with fluorescence imaging. [34] The primary outcome was the proportion of ulcers healed at 12 weeks by blinded assessment. Secondary outcomes included wound area reduction at 4 and 12 weeks and change in management decisions after fluorescence imaging. The proportion of ulcers healed at 12 weeks in the autofluorescence arm was 45% vs. 22% in the control arm. Wound area reduction was 40.4% (fluorescence positive) vs. 38.6% (control) at 4 weeks and 91.3% (fluorescence positive) vs. 72.8% (control) at 12 weeks. Wound debridement was the most common intervention in wounds with positive fluorescence imaging. There was a stepwise trend in healing favoring those with negative autofluorescence imaging, followed by those with positive autofluorescence

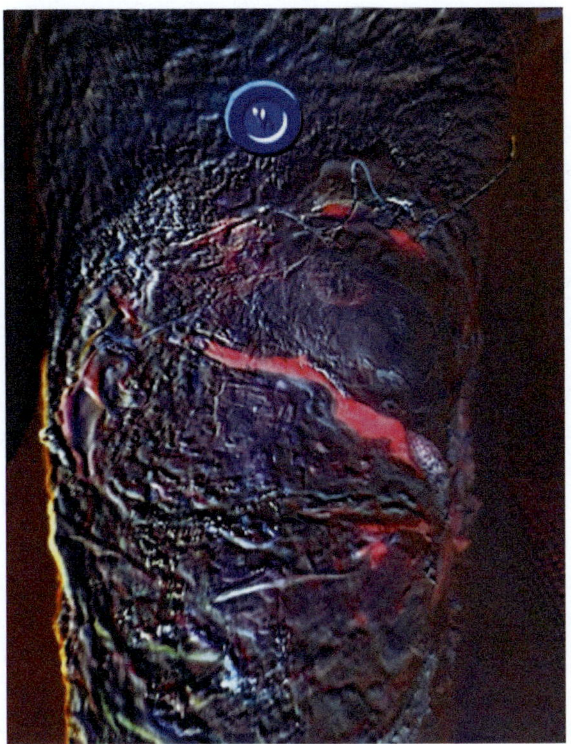

FIGURE 7.5 Failed CPT graft due to bacterial contamination. Bacterial fluorescence imaging can assess the wound bed before grafting and monitor the graft's evolution before significant contamination threatens its survival. (Photograph courtesy of Jose Ramirez GarciaLuna).

who had an intervention. The poorest outcome was the group with positive fluorescence and no corrective intervention.

Time to heal is important. Slow healing is associated with infection, hospitalization, and amputation. DFUs present for >30 days have fivefold increased odds of infection compared with those that heal, and infection increases the odds of hospitalization 55-fold and odds of amputation 154-fold when compared with noninfected DFUs. [35]

7.2.1.2 Practical Considerations

The ultimate utility of bacterial fluorescence imaging is treatment plan guidance. An example is differentiating between colonization (no host reaction) and infection (host reaction). Confusing colonization with infection can lead to spurious associations that may lead to expensive, ineffective, and time-consuming interventions. Some scenarios can be challenging. For example, immunocompromised patients do not show signs of infection as normal patients. Neutropenic patients (<500 neutrophils/mm^3) show no pyuria, no purulent sputum, little infiltrate, and no large consolidation on chest X-ray.

Using bacterial fluorescence imaging for qualitative assessment faces several challenges. Firstly, the boundary between colonization and infection is subject to various definitions. Different authors introduce various definitions, ranging from 10^4 CFU / g to 10^5 CFU / g. Researchers have proposed that there is a critical point between colonization and infection, termed critical colonization [36]. However, it is more of a theoretical concept at this point, which has yet to be established. One must also consider the specific "type" of bacteria, as some strains are more virulent than others. For example, Group A *Streptococcus pyogenes* is more virulent than *Streptococcus viridans*. In other words, 10^3 CFU/g of *Streptococcus pyogenes* is more severe than 10^4 CFU/g of *Streptococcus viridans*.

Secondly, it is difficult to establish the direct correlation between the fluorescence intensity and the bacterial load (bioburden). As both pyoverdine and porphyrins are by-products of bacterial presence, there is no simple relationship between their concentration and bioburden, as numerous factors can impact their production and fluorescence. For example, pyoverdine fluorescence is quenched in the presence of Fe^{2+} and Fe^{3+} [37].

While using bacterial fluorescence imaging, good imaging practices are necessary, such as cleaning the wound, removing as much blood as possible, and removing imaging artifacts from white bedsheets and gauze bandages when possible. Blood, in particular, can absorb the violet excitation light and mask other fluorescence signatures if not removed.

In addition, it should be considered that the violet excitation light cannot penetrate more than 1 mm to 1.5 mm into the skin. While this enables the detection of some subsurface bacteria, the presence of bacteria deeper within the wound tissue may not be visible, including deep tunneling infection.

To interpret bacterial fluorescent imaging properly, the user needs to understand anatomic sources of autofluorescence. The color of the fluorescent emission is a function of the Stokes shift. Connective tissue (collagen, elastin, fibrin), melanin, tendon, and bone are all fluorescent, with light emissions spanning a broad range of colors in the visible spectrum, between 420–700 nm. Some of these wavelengths can be removed from the image using custom emission filters (e.g., bandpass filters). For example, only fluorescence with wavelengths between 501–542 nm (cyan and green) and wavelengths between 601–664 nm (orange and red) can pass through to the camera. The emission filter also removes non-absorbed (reflected) excitation light from the image.

7.2.2 Metabolic Fluorescence Imaging

Metabolic fluorescence imaging or optical metabolic imaging (OMI) is a noninvasive, high-resolution, quantitative tool for monitoring cellular metabolism. It refers to a group of technologies that rely on measurements of the fluorescence intensity of NAD(P)H and FAD.

NADH and FAD are the principal electron donors and acceptors of oxidative phosphorylation, respectively. NADH has excitation maxima at 290 and 351 nm and emission maxima at 440 and 460 nm, while FAD has a single excitation maximum at 450 nm with its corresponding emission maxima at 535 nm.

Metabolic fluorescence imaging is based on the observation that the ratio of these fluorophores (NADH/FAD), called redox ratio, acts as a quantitative marker of the mitochondrial redox state of the tissue [38]. Here, redox (reduction-oxidation) refers to a type of chemical reaction in which the oxidation states of the substrate change.

In addition to the original redox scanner developed in the 70s by Britton Chance [39], several other technologies have been developed since then, including epifluorescence microscopy, confocal microscopy [40], two-photon excitation fluorescence microscopy (TPEF) [41], and fluorescence lifetime microscopy (FLIM) [42]. As such, they enabled a noninvasive 3D visualization of cell and tissue redox states. They have numerous applications in the metabolic monitoring of cells and tissues in cancer, neuroscience, and tissue engineering. For example, in [43], the technology was used to quantify the differences between the redox state of wounds in diabetic mice and the control mice. A significant correlation was observed between the redox state and the area of the wounds.

However, despite significant advances in recent years, these technologies are microscopy based and still confined to research labs. For the current state of this technology, readers can refer to recent reviews (see, e.g., [44]).

7.2.3 Endogenous Fluorescence: Other Fluorophores

Wood's lamp is probably the oldest known application of fluorescence in medicine. It is used to detect the fluorescence in skin and hair, which is a feature of some dermatophytes. It also can detect disorders of skin pigment, including vitiligo and other skin irregularities. The diagnostic utility of Wood's lamp is based on findings that certain bacteria, fungi, or changes in pigmentation of patient skin will cause the affected area of the skin to fluoresce; hence, this is used for inspection and detection of certain fungal infections [45]. It is also used to select infected skin and hairs for laboratory investigation.

All surface tissues (skin and mucosa) demonstrate autofluorescence, for example, from elastin and collagen. Thus, numerous endoscopic techniques exploit the fact that tumor cells alter the concentration of chromophores and fluorophores in tissues, promoting changes in the optical properties of the mucosa and, thus, altering the autofluorescence signal. For example, excitation at a wavelength of 400–460 nm results in the excitation of endogenous fluorophores of the healthy mucosa and the emission of a green light. The neoplastic mucosa exhibits a dark appearance because of a change in the quantity and quality of natural chromophores and fluorophores in tumors.

This abnormality is due to several mechanisms: (1) increase in the nuclear-cytoplasmic ratio, which consequently determines decreased autofluorescence as nuclei show no autofluorescence as compared with cytoplasm; (2) loss of collagen, submucosal collagen is the strongest fluorophore, which disappears due to thickening of the mucosa; and (3) neovascularization, inducing increased hemoglobin concentration, which absorbs autofluorescence light [46].

In addition, the tumor environment demonstrates red fluorescence, most likely caused by porphyrin accumulation. Thus, the decrease in green fluorescence and the increase in red fluorescence can indicate neoplastic transformation.

This approach (sometimes including also FAD and NADH autofluorescence) is used across various types of cancers *in vivo* (oral [47], skin [48]) and *ex vivo* (head and neck [49], oral [50], colon [51], and other types of cancers). *In vivo* endoscopic approaches will be discussed in more detail in section 11.3.

There are other approaches to autofluorescence-based cancer detection. However, they require excitation in the UV range of the spectrum and are more suitable for *ex vivo* measurements. For example, a more acidic tumor environment influences the fluorescence of pH-sensitive molecules, which could be used as a cancer biomarker. In particular, the serotonin emission maximum is shifted from 330 to 550 nm without a change in its absorption spectrum when the pH changes from neutral to strongly acidic (pH decreases) [52].

Tryptophan autofluorescence can also be used for tumor characterization. It is known that aggressive cancer cells have more large amino acid transporters on the cell membrane, which can increase the uptake of tryptophan from the surrounding environment. Zhang et al. [53] associated the fluorescence intensity ratio at 340 nm to the intensity at 460 nm (excitation at 300 nm) with the aggressiveness of the breast cancer cells.

The use of autofluorescence techniques in oncology was reviewed by Campbell et al. [54].

7.3 OTHER APPROACHES

In a broad sense, fluorescence imaging may include bioluminescent imaging (BLI)[2], which is invaluable in the laboratory and preclinical realms [55]. However, most BLI requires substrates that are activated for luminescence. Consequently, they are considered drugs in clinical development. As such, most BLIs are restricted to the preclinical area.

However, some substances are approved by regulators for clinical use for specific clinical indications. Thus, numerous off-label uses were tried for them, particularly 5-aminolevulinic acid (ALA). ALA is a naturally occurring amino acid that is a protoporphyrin IX (PpIX) precursor, a heme precursor in the biosynthetic pathway. After exogenous administration of ALA, PpIX accumulates mainly in cancer cells owing to the impaired metabolism of ALA to PpIX in mitochondria, which results in a red fluorescence following irradiation with blue light [56]. Fluorescence navigation using ALA has been tried for evaluating both the surgical resection margins and the extension of the lesion in gastric cancer [56]. In addition, ALA can be used to detect peritoneal metastases during preoperative staging laparoscopy [57] and evaluate lymph node metastases [58]. The clear advantage of ALA/PpIX is their theranostic capability, as they can be used in photodynamic diagnostics (PDD) and photodynamic therapy (PDT) by releasing singlet oxygen.

7.4 FLUORESCENCE-BASED TECHNIQUES

Fluorescence can be observed using a relatively simple setup. In the simplest case, it may require only an emission filter to filter out strong excitation light that masks a weak fluorescence signal. This relatively simple steady-state approach is used in existing clinical modalities like bacterial imaging or fluorescence angiography. However, in addition to this, multiple fluorescence imaging modalities have emerged for biomedical applications. They allow single-molecule detection and below-diffraction limit resolution (for example, near-field imaging). Below, we consider several examples that are widely used in biomedical research and clinical practice. However, this list is definitely not exhaustive. For example, endoscopic applications will be considered in Chapter 11 (Endoscopy). Similarly, two-photon fluorescence will be considered in section 12.2.1.2.

7.4.1 FRET

Förster resonance energy transfer (FRET) is a non-radiative transfer of the optical excitation from an excited electronic level of one molecule, called the donor, to a resonant electronic level of another molecule, called the acceptor. If the electronic levels of donor (D) and acceptor (A) are in a strong resonance, the optical excitation can be up to 10 nm. Energy transfer efficiency exponentially decays as a sixth power of D-A separation distance. As such, FRET is an indicator of the collocation of donor and acceptor molecules.

FRET is an invaluable molecular diagnostics tool with multiple applications in clinical practice and biomedical research. It can be used to quantify almost any target of interest (proteins, nucleic acids, metabolites, drugs, toxins, human cells, microbes, and other pathogens) from various types of clinical specimens (body fluids, cells, tissues).

According to Qui et al. [59], FRET's applications in clinical diagnostics include the following:

> "Simple and rapid molecular diagnostics via homogeneous immunoassays;
> High throughput molecular diagnostics via microarrays;
> Highly sensitive molecular diagnostics via real-time PCR;
> Molecular dynamic diagnostics via live cell imaging."

FRET-based molecular diagnostic assays/platforms are key techniques in genetic diagnostics of diseases and infectious pathogens, cancer risk prediction, disease progression and early diagnostics, and prenatal screening. For example, in oncology, it can be used to detect oncogenes [60], miRNAs [61], protein cancer biomarkers [62], and protein–protein interactions [63].

Given the simplicity of sample preparation, this technology can be used in POC settings.

7.4.2 Time-Resolved Fluorescence

Time-resolved fluorescence (TRF) is a group of technologies that capitalize on the extra property of the fluorescence lifetime. Medical applications of fluorescence lifetime techniques were reviewed by Marcu [64].

7.4.2.1 Time-Resolved Detection

Time-resolved detection (TRD) is a technique to improve signal-to-noise ratio during exogenous fluorescence imaging by rejecting tissue auto-fluorescence. TRD is based on the difference in fluorescence lifetimes between the endogenous and specially selected exogenous fluorophores. It is achieved by separating excitation and signal collection in time to allow fast fluorescence from native fluorophores (typically, on the scale of single nanoseconds or faster) to die out. In the typical scenario, it is used in conjunction with phosphorescent bioprobes like transition metal or lanthanide complexes or nanoparticles. For example, the most commonly used lanthanide chelate label is the europium ion (Eu^{3+}).

7.4.2.2 FLIM

Fluorescence lifetime imaging microscopy (FLIM) goes further than the TRD approach and aims to measure radiation decay (as a proxy of fluorescence lifetime) as a unique signature of a fluorophore.

FLIM is not sensitive to the concentration of fluorophores. Instead, it measures their presence. FLIM has several advantages over intensity-based methods. For example, FLIM is a self-referenced measurement (i.e., independent of absolute detected intensity), so FLIM experiments do not require the throughput calibration steps needed for intensity-based experiments. Additionally, as the fluorescence lifetime is extremely sensitive to changes in the chemical composition or even environment, the fluorescence lifetime can be used to discriminate between highly similar structures. In particular, it can be used to differentiate between free and protein-bound states of NAD(P)H and FAD. The fluorescence lifetime of NAD(P)H is significantly shorter in the free state (~400 ps) compared with the protein-bound state (~1 to 5 ns) of the molecule [65]. Conversely, FAD has a longer lifetime in its free state (2.3 to 2.9 ns) compared with its protein-bound state (<0.1 ns). As such, FLIM has been extensively used in autofluorescent molecular imaging to study cellular metabolism.

In addition to label-free (endogenous) applications, FLIM can be used for exogenous fluorescence. In particular, numerous optical probes have been developed for both *in vivo* and *in vitro* applications to capitalize on the sensitivity of FLIM to microenvironmental parameters, including viscosity, temperature, acidity, and oxygenation.

There are two primary approaches: time domain (TD-FLIM) and frequency domain (FD-FLIM). TD-FLIM uses a short pulse of light (short relative to the fluorescence lifetime of the fluorophore) for excitation and then records the exponential decay of fluorescent molecules either directly (i.e., by gated detection or pulse sampling) or by using time-resolved electronics that bin photons by their arrival times (time-correlated single-photon counting, or TCSPC). As such, it relies on pulsed lasers.

FD-FLIM analyzes the phase and amplitude of fluorescence signals modulated at different frequencies. A significant advantage of frequency-domain FLIM over time-domain FLIM techniques, such as TCSPC, is acquisition speed, making frequency-domain an ideal technique for measuring rapid cellular events.

FLIM can be performed using various imaging modalities, such as laser scanning microscopy (LSM) and wide-field illumination (WFI) microscopy. FLIM is often combined with FRET, confocal and multiphoton microscopy, and OCT.

FLIM has multiple applications in medicine, particularly in surgical oncology. In particular, FLIM is well suited for the intraoperative delineation of tumors based on rapid evaluation of the extent of molecular changes (neoplastic area). It is used for brain and breast cancer surgeries. It has been translated into clinical practice in oral cavity and oropharyngeal cancers by integrating transoral robotic surgical platforms (TORS) to detect surgical margins.

A comprehensive review of FLIM techniques was provided by Datta et al. [66]. Clinical applications of FLIM have been reviewed by Alfonso-Garcia et al. [67].

7.5 CONCLUSIONS

Luminescence is the process when the exited absorption center (e.g., molecule) returns to its ground state through photon emission. Depending on fluorescence lifetime, luminescence is split into fluorescence (short lifetime on a scale of nanoseconds or less) and phosphorescence (longer lifetime on a scale from 10^{-4} seconds to minutes or even hours).

Fluorescence is characterized by the emission and excitation spectra. The emission wavelength is typically longer than the excitation wavelength (Stokes shift).

Based on the origin of the fluorophores, fluorescence can be split into endogenous (or autofluorescence) and exogenous.

Fluorescence imaging is a powerful tool in life sciences. In addition, it has multiple applications in medical practice. Medical applications use both endogenous and exogenous fluorescence.

Native fluorophores in human skin include collagen, elastin, amino acids (tryptophan and tyrosine), NADH, NAD, FAD, and porphyrins. In addition, certain bacteria produce fluorophores as their their metabolites. Thus, endogenous fluorescence imaging has applications in metabolic imaging, oncology, and wound care.

In addition to endogenous fluorescence imaging, multiple exogenous fluorescence imaging techniques emerged. In these techniques, the dye is typically injected into the bloodstream. Several dyes (for example, fluorescein and indocyanine green, ICG) are approved for clinical use.

The popularity of fluorescence-based techniques is caused by their relatively simple setup. However, multiple sophisticated fluorescence-based techniques like multiphoton excitation, FRET, and TRF have emerged in recent years.

NOTES

1 CFU stands for a colony forming unit.
2 Bioluminescent imaging is based on the detection of visible light produced by living organisms during enzyme (luciferase)-mediated oxidation of a molecular substrate when the enzyme is expressed *in vivo* as a molecular reporter.

REFERENCES

1. NA Wisniewski, SP Nichols, SJ Gamsey, et al. "Tissue-integrating oxygen sensors: Continuous tracking of tissue hypoxia". *Adv Exp Med Biol* 977: 377–383 (2017).
2. SP Nichols, MK Balaconis, RM Gant, et al. "Long-term in vivo oxygen sensors for peripheral artery disease monitoring". *Adv Exp Med Biol* 1072: 351–356 (2018).
3. MP Coogan, V Fernandez-Moreira. "Progress with, and prospects for, metal complexes in cell imaging". *Chem Commun* 50: 384–399 (2014).
4. Z Jendželovská, R Jendželovský, B Kuchárová, P Fedoročko. "Hypericin in the light and in the dark: Two sides of the same coin". *Front Plant Sci* 7: 560 (2016).
5. JMV Nguyen, L Hogen, S Laframboise, et al. "The use of indocyanine green fluorescence angiography to assess anastomotic perfusion following bowel resection in surgery for gynecologic malignancies – a report of 100 consecutive anastomoses". *Gynecol Oncol* 158(2): 402–406 (2020).

6. K Igari, T Kudo, H Uchiyama, et al. "Indocyanine Green angiography for the diagnosis of peripheral arterial disease with isolated infrapopliteal lesions". *Ann Vasc Surg* 28(6): 1479–1484 (2014).

7. Y Woo, S Chaurasiya, M O'Leary, et al. "Fluorescent imaging for cancer therapy and cancer gene therapy". *Mol Ther Oncolytics* 23: 231–238 (2021).

8. H Maeda, H Nakamura, J Fang. "The EPR effect for macromolecular drug delivery to solid tumors: Improvement of tumor uptake, lowering of systemic toxicity, and distinct tumor imaging in vivo". *Adv Drug Deliv Rev* 65: 71–79 (2013).

9. N Patel, M Allen, K Arianpour, R Keidan. "The utility of ICG fluorescence for sentinel lymph node identification in head and neck melanoma". *Am J Otolar* 42(5): 103147 (2021).

10. S Gentileschi, M Servillo, R Albanese, et al. "Lymphatic mapping of the upper limb with lymphedema before lymphatic supermicrosurgery by mirroring of the healthy limb". *Microsurgery* 37: 881–889 (2017).

11. M Coelho, N Maghelli, IM Tolić-Nørrelykke. "Single-molecule imaging in vivo: The dancing building blocks of the cell". *Integrative Biol* 5(5): 748–758 (2013).

12. G Donnert, J Keller, R Medda, et al. "Macromolecular-scale resolution in biological fluorescence microscopy". *PNAS* 103: 11440–11445 (2006).

13. JH Samies, M Gehling, TE Serena, RA Yaakov. "Use of a fluorescence angiography system in assessment of lower extremity ulcers in patients with peripheral arterial disease: A review and a look forward". *Seminars Vasc Surg* 28(3–4): 190–194 (2015).

14. A Salleh, MB Fauzi. "The in vivo, in vitro and in ovo evaluation of quantum dots in wound healing: A review". *Polymers* 13: 191 (2021).

15. KT Yong, WC Law, R Hu, et al. "Nanotoxicity assessment of quantum dots: From cellular to primate studies". *Chem Soc Rev* 42: 1236–1250 (2013).

16. X Xu, R Ray, Y Gu, et al. "Electrophoretic analysis and purification of fluorescent single-walled carbon nanotube fragments". *J Am Chem Soc* 126(40): 12736–12737 (2004).

17. M Pirsaheb, S Mohammadi, A Salimi, et al. "Functionalized fluorescent carbon nanostructures for targeted imaging of cancer cells: A review". *Microchim Acta* 186: 231 (2019).

18. J-CG Bünzli. "Lanthanide light for biology and medical diagnosis". *J Lumines* 170(3): 866–878 (2016).

19. J.-CG Bünzli, *Luminescence in Lanthanide Coordination Compounds and Nanomaterials*, A de Bettencourt-Dias, Ed. Oxford: Wiley-Blackwell, p. 125 (2014).

20. J.-CG Bünzli. "Lanthanide luminescence for biomedical analyses and imaging". *Chem Rev* 110: 2729 (2010).

21. W Zheng, D Tu, P Huang, et al. "Time-resolved luminescent biosensing based on inorganic lanthanide-doped nanoprobes". *Chem Commun* 51: 4129 (2015).

22. B Kjeldstad, T Christensen, A Johnsson. "Porphyrin photosensitization of bacteria". *Adv Exp Med Biol* 193: 155–159 (1985).

23. YS Cody, DC Gross. "Characterization of pyoverdin (pss), the fluorescent siderophore produced by Pseudomonas syringae pv. Syringae". *Appl Environ Microbiol* 53: 928–934 (1987).

24. L Le, M Baer, P Briggs, et al. "Diagnostic accuracy of point-of-care fluorescence imaging for the detection of bacterial Burden in wounds: Results from the 350-patient FLAAG trial". *Adv Wound Care* 10(3): 123–136 (2021).

25. R Hill, KY Woo. "A prospective multi-site observational study incorporating bacterial fluorescence information into the UPPER/LOWER wound infection checklists". *Wounds* 32: 299–308 (2020).

26. TE Serena, K Harrell, L Serena, RA Yaakov. "Real-time bacterial fluorescence imaging accurately identifies wounds with moderate-to-heavy bacterial burden". *J Wound Care* 28: 346–357 (2019).

27. K Ottolino-Perry, E Chamma, KM Blackmore, et al. "Improved detection of clinically relevant wound bacteria using autofluorescence image-guided sampling in diabetic foot ulcers". *Int Wound J* 14: 833–841 (2017).

28. LM Jones, D Dunham, MY Rennie, et al. "In vitro detection of porphyrin-producing wound bacteria with real-time fluorescence imaging". *Future Microbiol* 15: 319–332 (2020).

29. N Farhan, S Jeffery, "Diagnosing burn wounds infection: The practice gap & advances with MolecuLight bacterial imaging". *Diagnostics*, 11: 268 (2021).

30. R Raizman, D Dunham, L Lindvere-Teene, et al. "Use of a bacterial fluorescence imaging device: Wound measurement, bacterial detection and targeted debridement". *J Wound Care* 28: 824–834 (2019).

31. R Hill, MY Rennie, J Douglas. "Using bacterial fluorescence imaging and antimicrobial stewardship to guide wound management practices: A case series". *Ostomy Wound Manag* 64: 18–28 (2018).

32. R Raizman. "Fluorescence imaging guided dressing change frequency during negative pressure wound therapy: A case series". *J Wound Care* 28: S28–S37 (2019).

33. B Aung. "Can fluorescence imaging predict the success of CTPs for wound closure and save costs?" *Today's Wound Clin* 13: 22–25 (2019).

34. S Rahm, J Woods, S Brown, et al. "The use of point-of-care bacterial auto-fluorescence imaging in the management of diabetic foot ulcers: A pilot randomized controlled trial". *Diab Care* 45(7): 1601–1609 (2022).

35. LA Lavery, DG Armstrong, RP Wunderlich, et al. "Risk factors for foot infections in individuals with diabetes". *Diab Care* 29: 1288–1293 (2006).

36. RJ White, KF Cutting. "Critical colonization – the concept under scrutiny". *Ostomy Wound Manage* 52(11): 50–56 (2006).

37. MF Yoder, WS Kisaalita. "Fluorescence of Pyoverdin in response to iron and other common well water metals". *J Env Sci Health, Part A* 41(3): 369–380 (2006).

38. MF la Cour, S Mehrvar, JS Heisner, et al. "Optical metabolic imaging of irradiated rat heart exposed to ischemia-reperfusion injury". *J Biomed Opt* 23(1): 1–9 (2018).

39. B Chance, B Schoener, R Oshino, et al. "Oxidation-reduction ratio studies of mitochondria in freeze-trapped samples. NADH and flavoprotein fluorescence signals". *J Biol Chem* 254(11): 4764–4771 (1979).

40. BR Masters, A Kriete, J Kukulies. "Ultraviolet confocal fluorescence microscopy of the in vitro cornea: Redox metabolic imaging". *Appl Opt* 32: 592–596 (1993).

41. SH Huang, AA Heikal, WW Webb. "Two-photon fluorescence spectroscopy and microscopy of NAD(P)H and flavoprotein". *Biophys J* 82: 2811–2825 (2002).

42. TS Blacker, ZF Mann, JE Gale, et al. "Separating NADH and NADPH fluorescence in live cells and tissues using FLIM". *Nat Commun* 5: 3936 (2014).

43. S Mehrvar, KT Rymut, FH Foomani, et al. "Fluorescence imaging of mitochondrial redox state to assess diabetic wounds". *IEEE J Transl Eng Health Med* 18(7): 1800809 (2019).

44. OI Kolenc, KP Quinn. "Evaluating cell metabolism through autofluorescence imaging of NAD(P)H and FAD". *Antioxid Redox Signal* 30(6): 875–889 (2019).

45. DM Al Aboud, W Gossman. "Wood's light". In *StatPearls [Internet]*. Treasure Island (FL): StatPearls Publishing (2024). www.ncbi.nlm.nih.gov/books/NBK537193/

46. K Ragunath. "Autofluorescence endoscopy--not much gain after all?" *Endoscopy* 39: 1021–1022 (2007).

47. DM Roblyer, C Kurachi, V Stepanek, et al. "Comparison of multispectral wide-field optical imaging modalities to maximize image contrast for objective discrimination of oral neoplasia". *J Biomed Opt* 15(6): 066017 (2010).

48. RA Romano, RG Teixeira Rosa, AG Salvio, et al. "Multispectral autofluorescence dermoscope for skin lesion assessment". *Photodiagnosis Photodyn Ther* 30: 101704 (2020).

49. M Halicek, JD Dormer, JV Little, et al. "Tumor detection of the thyroid and salivary glands using hyperspectral imaging and deep learning". *Biomed Opt Express* 11: 1383–1400 (2020).

50. Y-J Yan, N-L Cheng, C-I Jan, et al. "Band-selection of a portal LED-induced autofluorescence multispectral imager to improve oral cancer detection". *Sensors* 21: 3219 (2021).

51. J Deal, S Mayes, C Browning, et al. "Identifying molecular contributors to autofluorescence of neoplastic and normal colon sections using excitation-scanning hyperspectral imaging". *J Biomed Opt* 24(2): 021207 (2018).

52. K Berland. "Basics of fluorescence". In *Methods in Cellular Imaging*. New York, NY: Springer (2001).

53. L Zhang, Y Pu, J Xue, et al. "Tryptophan as the fingerprint for distinguishing aggressiveness among breast cancer cell lines using native fluorescence spectroscopy". *J Biomed Opt* 19(3): 037005 (2014).

54. JM Campbell, A Habibalahi, S Handley, et al. "Emerging clinical applications in oncology for non-invasive multi- and hyperspectral imaging of cell and tissue autofluorescence". *J Biophotonics* 16(9): e202300105 (2023).

55. RT Sadikot, TS Blackwell. "Bioluminescence imaging". *Proc Am Thorac Soc* 2(6): 511–512, 537–540 (2005).
56. T Namikawa, T Yatabe, K Inoue, et al. "Clinical applications of 5-aminolevulinic acid-mediated fluorescence for gastric cancer". *World J Gastroenterol* 21(29): 8769–8775 (2015).
57. K Kishi, Y Fujiwara, M Yano, et al. "Staging laparoscopy using ALA-mediated photodynamic diagnosis improves the detection of peritoneal metastases in advanced gastric cancer". *J Surg Oncol* 106: 294–298 (2012).
58. N Koizumi, Y Harada, Y Murayama, et al. "Detection of metastatic lymph nodes using 5-aminolevulinic acid in patients with gastric cancer". *Ann Surg Oncol* 20: 3541–3548 (2013).
59. X Qiu, N Hildebrandt. "A clinical role for Förster resonance energy transfer in molecular diagnostics of disease". *Expert Rev Mol Diagn* 19(9): 767–771 (2019).
60. SA Belinsky, HE Carraway, VJ Bailey, et al. "MS-qFRET: A quantum dot-based method for analysis of DNA methylation". *Genome Res* 19: 1455–1461 (2009).
61. D Gustafson, K Tyryshkin, N Renwick. "microRNA-guided diagnostics in clinical samples". *Best Pract Res Clin Endocrinol Metab* 30(5): 563–575 (2016).
62. MJ Chen, YS Wu, GF Lin, et al. "Quantum-dot-based homogeneous time-resolved fluoroimmunoassay of alpha-fetoprotein". *Anal Chim Acta* 34: 100–105 (2012).
63. G Weitsman, PR Barber, LK Nguyen, et al. "HER2-HER3 dimer quantification by FLIM-FRET predicts breast cancer metastatic relapse independently of HER2 IHC status". *Oncotarget* 7: 51012–51026 (2014).
64. L Marcu. "Fluorescence lifetime techniques in medical applications". *Ann Biomed Eng* 40(2): 304–331 (2012).
65. JR Lakowicz, H Szmacinski, K Nowaczyk, et al. "Fluorescence lifetime imaging". *Anal Biochem* 202(2): 316–330 (1992).
66. R Datta, TM Heaster, JT Sharick, et al. "Fluorescence lifetime imaging microscopy: Fundamentals and advances in instrumentation, analysis, and applications". *J Biomed Opt* 25(7): 1–43 (2020).
67. A Alfonso-Garcia, J Bec, B Weyers, et al. "Mesoscopic fluorescence lifetime imaging: Fundamental principles, clinical applications and future directions". *J Biophotonics* 14(6): e202000472 (2021).

CHAPTER 8

Thermographic Physiological Optical Imaging Techniques

THERMOGRAPHY IS A PASSIVE modality, which has been known in medicine for a long time. With technological advances, thermographic techniques have been adopted into clinical practice in recent years.

Temperature measurement modalities in medicine can be categorized across several dimensions [1]: a) contact vs. noncontact and b) single (or several) point vs. multipoint measurements.

The primary utility of thermography as a diagnostic optical imaging modality is large-area thermographic imaging, which can reveal spatial information. Even though temperature distribution in certain locations (e.g., soles) can be measured by contact modalities (e.g., temperature monitoring mat and temperature monitoring wearables [2]), noncontact imaging has multiple advantages over other thermographic approaches. As such, this chapter will focus on thermal camera applications.

Noncontact (remote) temperature measurement techniques do not measure the temperature directly. Instead, they measure energy flow from the object and derive temperature from these measurements (radiometry). Radiometry is based on the fact that all objects with temperatures above 0K emit electromagnetic radiation in a broad range of wavelengths. However, most radiation from surrounding objects (e.g., in the 0–100°C temperature

DOI: 10.1201/9781003482505-10

range) comes from wavelengths three micrometers and longer, commonly referred to as thermal infrared. For example, the radiation peak from an object at 300K (27°C) is 9 µm. Objects in this temperature range do not emit noticeable radiation in the visible range of the spectrum. Only objects heated 500°C and above emit noticeable radiation in the visible range of the spectrum (i.e., red-hot metal).

Thermal imaging sensors operate in the middle (MWIR, 3–5 µm) or long (LWIR, 8–14 µm) wavelength infrared (IR) spectral ranges. As such, non-contact thermography is often referred to as infrared thermography or IRT.

The most common materials for thermosensors are amorphous Si (a-Si), InSb, InGaAs, HgCdTe, and quantum well-integrated photodetectors (QWIP) arranged into focal plane arrays or FPA. While room temperature operations are sufficient for many applications, the cryogenic cooling of IR detectors is required for specific applications to achieve high performance. Uncooled detectors are mainly based on pyroelectric and ferroelectric materials or microbolometer technology. Thermal imaging sensors for biomedical research and clinical applications use low-cost, uncooled FPA microbolometers in the LWIR range. The resolution of thermographic sensors is much lower than regular RGB cameras. Currently, it is in the range of 120 × 160 for standard applications and 640 x 480 for high-end applications.

8.1 CLINICAL UTILITY

Vincent Czerny developed the first thermographic instrument (Evaporograph) in 1925 and documented the first infrared image of a human subject in Frankfurt in 1928 [3]. The medical use of infrared thermography started in 1952 in Germany when Schwamm and Reeh developed a single-detector infrared bolometer for sequential thermal measurement of defined regions of the human body surface for diagnostic purposes [4].

Thermography in a medical context measures temperature distribution in superficial tissues. The skin temperature is affected by multiple factors, including environmental factors. However, in thermal equilibrium, it depends primarily on the blood supply. As such, thermography can be used to detect abnormalities in blood circulation. Two primary cases of practical importance are:

- Decreased blood supply, which causes decreased skin temperature

- Elevated blood supply, which causes increased skin temperature

Thermography has been studied extensively across different medical domains. Most widely, it is used for fever screening. However, it is also used in other domains, including wound care and oncology. Lahiri et al.[5] reviewed medical applications of infrared thermography across various medical domains. There are also narrower reviews of applications of IRT in wound care, surgery, and sports medicine [1] or ophthalmology [6].

8.1.1 Wound Care

The utility of thermography in wound care has been known since the early 1960s when Lawson et al. [7] used infrared scanning to predict burn depth with an accuracy of 90%, as confirmed by histology.

Thermography has significant potential as an adjuvant technique in wound care assessment. For example, elevated temperature is a reliable marker of inflammation and can thus predict the risk of ulceration, infection, and amputation [8]. Similarly, the decreased temperature may be a sign of insufficient blood supply and indicate ischemia. The applications of thermography in wound care were reviewed in [1].

8.1.1.1 Diagnosis of PAD

Under controlled conditions, skin temperature directly reflects arterial blood flow, which transports oxygen, nutrients, and heat to the extremities [9]. The extremities' high surface-to-volume ratio means the extremities will cool if there is a decrease in blood flow. As such, various studies have investigated IRT's role as a diagnostic tool for detecting peripheral arterial disease (PAD) (see, for example, [10]) and as a tool for post-procedural assessment/surveillance of angioplasties (see, for example, [11]).

8.1.1.2 Flap Perfusion

Thermography can be used in plastic surgery to identify and monitor arterial perfusion of free flaps. [12]

A prerequisite to flap survival is the identification of the dominant perforator artery that supplies the flap. A "freestyle" approach based on general anatomy is often used to design these flaps, as conventional imaging techniques for specific perforator identification may be too expensive or unavailable. However, unexpected variations in anatomy can lead to flap failure. Using a thermal imaging camera is a cheaper and more universal means to identify the requisite perforators upon which a free flap can be designed and monitored.

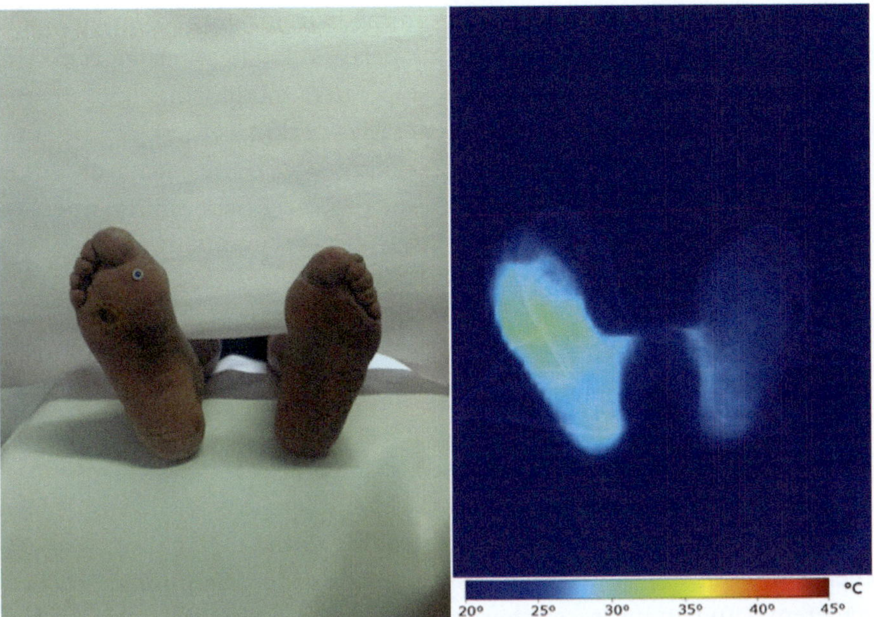

FIGURE 8.1 Thermographic assessment of peripheral arterial disease (PAD). Limbs with PAD show significantly lower temperatures than the contralateral extremity. In this case, a diabetic individual with PAD on the left foot is presented. A screen is used to thermally isolate the rest of the body, showing a stark contrast between the feet's temperature. (Photograph courtesy of Dr. Jose Ramirez-GarciaLuna).

In particular, thermography (including smartphone thermography) can be used preoperatively to identify arterial perforators or vascular "hot spots" within the desired donor site territory. Intraoperative selection of the arterial perforators and flap dissection can be facilitated using IRT as a real-time guide. In addition, intermittent postoperative monitoring using thermal images provides a comparison method to determine if the anastomotic perfusion is compromised.

A concordance study compared smartphone perforator detection with CT angiography. [13] Smartphone IRT was highly accurate, with a sensitivity of 100% and a specificity of 98%. Other recognized techniques for perforator identification include magnetic resonance angiography or color Duplex ultrasound, which are reliable alternatives but are expensive, time consuming, and not universally available. [14, 15]

In addition, thermography offers immediate insights to provide adequate intraoperative confirmation of perfusion. The thermogram also provides an additional means for postoperative monitoring. Surveillance is important in the immediate postoperative period as there can be complications such as arterial kinking, compression from edema or minor hematomas, etc. Early identification leads to early intervention and successful outcomes. A thermogram is a near-perfect surveillance tool, simple to obtain, noninvasive, and accurate.

8.1.1.3 Inflammation Detection

Inflammation is a complex physiological response triggered by tissue damage. Inflammation is characterized by a series of events aimed at removing the injurious agent and initiating tissue repair. One of the earliest responses includes the induction of hyperemia, which refers to a localized increase in the blood flow to the affected area, leading to the cardinal signs of erythema and warmth [16]. The process by which acute inflammation induces hyperemia and an increased thermal load includes histamine, prostaglandins, and bradykinin-induced meta-arteriole vasodilation; increased vascular permeability, which also leads to vascular fluid leakage and edema; and decreased vascular tone in the capillary bed, also leading to increased flow and plasma leakage [17].

Infrared thermography is an excellent tool for visualizing acute and chronic inflammatory changes [8]. For example, localized temperature increases after tissue injury were documented [18]. The authors followed a series of patients experiencing traumatic injuries to the extremities. They obtained time-lapsed thermal images showing dramatic increases in the affected areas' heat load and dissipation of the heat as time passed. Interestingly, in one case, the heat load followed the vascular distribution of the hand, demonstrating how flow and heat are intimately related during inflammation. Similarly, in chronic inflammatory conditions, localized temperature increases have been demonstrated to correlate with areas of inflammation diagnosed by other imaging modalities, such as ultrasound [19], or to clinical severity scoring systems [20].

Multiple approaches were proposed for the assessment of chronic wounds. In particular, diabetic foot ulceration risk can be assessed based on plantar temperature distributions. Benbow et al. [21] assessed the risk of ulceration and ischemic foot disease based on the mean foot temperature determined from eight standard sites on the plantar surface. They found that an elevated mean foot temperature was associated with an increased risk for neuropathic foot ulceration. Diabetic patients with normal or low mean foot temperature were at risk for ischemic foot disease.

Another approach is to compare temperature maps of the individual's contralateral (left foot to their right) sites (an asymmetry analysis). This method has the advantage of being specific to the patient. However, it is dependent on geometrical symmetry between feet [22]. Therefore, preventive care is recommended when a patient is observed with temperature asymmetry exceeding $2.2°C\,(4\,°F)$ for at least two consecutive days between contralateral sites [23]. Using a remote temperature monitoring mat and the 2.2°C asymmetry approach, Frykberg et al. [24] predicted 97% of all non-acute plantar DFU on average 35 days before clinical presentation with a specificity of 43%.

Nagase et al. [25] identified 20 patterns of thermal distribution to aid in diabetic foot assessment and surgical procedures. Bharara et al. [26] proposed a wound inflammatory index or temperature index for diabetic foot assessment, which is based on the difference in mean foot temperature and the wound bed, the area of the wound bed, and the area of the isotherm (highest or lowest temperature area).

Lavery et al. [27], in a multicenter study on 129 patients in remission, found that unilateral once-daily foot temperature monitoring can predict 91% of impending non-acute plantar foot ulcers on average 41 days before clinical presentation.

8.1.1.4 Infection Detection

Infection also produces a localized increase in temperature as part of the body's immune response, which is triggered by the presence of pathogens in the affected area. Similarly to that observed during inflammation, it is mainly driven by increased vascular permeability and blood flow [28]. Erythema, localized heat increases, and swelling are cardinal signs of wound infection; however, they are challenging to assess clinically. A study that compared the rate of clinical vs. histopathology-confirmed wound infection demonstrated that the accuracy of clinical inspection by trained observers is less than 70% [29]. To mitigate these shortcomings, multiple clinical decision rules have been developed to diagnose wound infections clinically, including the NERDS and STONEES mnemonics. While these rules offer a more objective assessment of a wound's characteristics, their sensitivities and specificities remain moderate, hovering at around 70 and 80%, respectively [30]. A study by Woo and Sibbald [31] demonstrated that by combining any three clinical signs present in these rules, the specificity

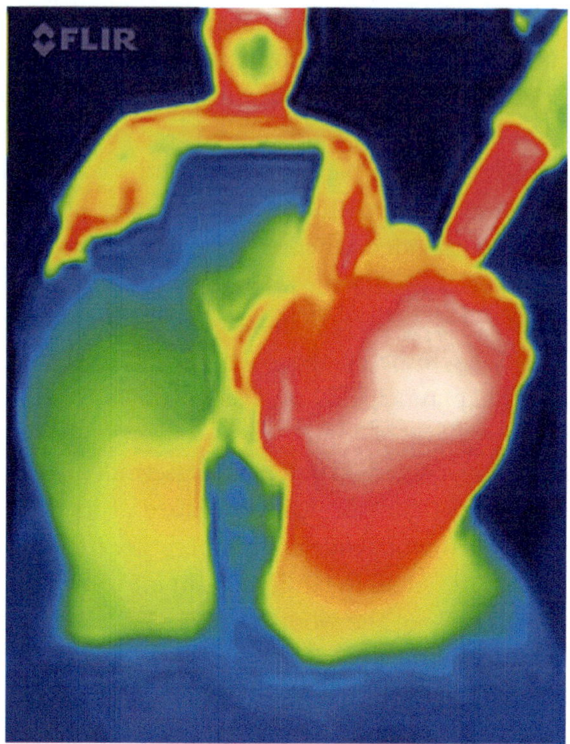

FIGURE 8.2 Thermographic assessment of Charcot arthropathy changes. Charcot arthropathy is characterized by the sudden onset of pain and inflammation in the foot of uncontrolled diabetic individuals. This disease is thermographically hallmarked by increased temperature in the complete foot, as opposed to a localized area in the case of infections. In this image, a significant increase in the temperature of the left foot can be observed. (Photograph courtesy of Dr. Jesus E Arriaga Caballero).

for infection detection rose to 90%, albeit with a reduction of the specificity to 70%. Interestingly, the same authors noted that the single highest predictive sign of infection was a localized temperature increase, with an odds ratio (OR) of 8.05.

For this reason, thermography has become an adequate tool for supporting the diagnosis of infection [8]. Rahbek et al. used this tool to discriminate between patients with orthopedic external pin infections and those without [32]. Their findings demonstrate that a cutoff value of 34°C for infection has a sensitivity of 73%, specificity of 67%, positive predictive value of 10%, and negative predictive value of 98%. Moreover, they also found an intrarater agreement for thermography of ICC 0.85 (0.77–0.92), demonstrating that thermography can be used very reproducibly by different observers, thus removing some of the variability and subjectivity of clinical inspection alone.

Similarly, a study by Chanmugan et al. [33] demonstrated a temperature gradient in patients with inflamed vs. infected wounds. Those patients whose wounds were inflamed but not deemed to be clinically infected had a thermal load of +1.5° C to 2.2° C, compared to those with infected wounds, which were in the range of +4° C to 5° C. Additionally, they were able to document how, in response to antibiotic therapies, the local temperature of the wounds dropped to +0.8° C to 1.1° C gradients, which were consistent with normal wounds. Another example of the use of thermography to detect inflammation and infection was provided by Oe et al. [34], who, in addition to detecting inflammation, found that thermography may be able to predict osteomyelitis; a severe complication of the diabetic wound before visible signs of infection are shown.

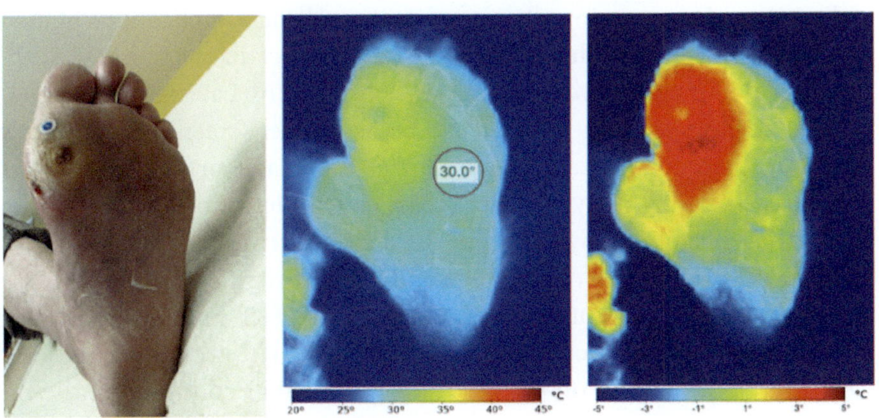

FIGURE 8.3 Thermographic assessment of a diabetic ulcer infection. Under thermal imaging, deep tissue infections appear as localized, intense "hotspots," usually above a thermal gradient of +2.5°C. In this case, the patient presented with an abscess around the first metatarsal area. Thermal imaging allowed its early identification and management. (Photograph courtesy of Dr. Jose Ramirez-GarciaLuna).

8.1.2 Oncology

Thermography has a long history in oncology. It is based on the observation that many tumors are hot, and the hotter the tumor, the worse the prognosis [35]. In particular, tumors generally have an increased blood supply and an increased metabolic rate, which leads to localized high-temperature spots over such areas, which can be visualized by IRT. For example, Gamagami [36] studied angiogenesis in breast cancer by IRT and reported that in 15% of cases IRT could detect cancers that were not noticeable by mammography.

Anbar [37] identified that

> abnormal behavior of skin temperature can be manifest in two principal modalities: (1) pathological changes in the spatial distribution of temperature over the skin surface, (2) pathological changes in the dynamic temperature behavior, i.e., warming, cooling, or periodic cooling of a given sub area of skin.

While diagnostic thermography was tried for many cancers like malignant melanoma [38], its primary utility in oncology is in breast cancer screening.

Breast cancer screening was one of the first clinical applications of thermography. Thermography entered clinical practice in the late 50s as the first thermal image of breast cancer was obtained by Lawson [39] in 1956. Since then, it was rapidly adopted as a breast cancer screening technology. However, its acceptance rapidly ended in 1977 after a report by Feig et al. [40] compared thermography to other breast cancer detection methods. They found that thermography came out third after ultrasound and mammography, with a sensitivity of only 39% and a specificity of 82% [40].

However, more recent reviews that compare IRT and other imaging techniques for breast screening conclude that IRT provides additional functional information on the thermal and vascular condition of the tissues [41]. For example, a prospective clinical study published by Arora et al. [42] evaluated 92 women with a previous suspicious mammogram or ultrasound result and demonstrated the precision and quality of the thermographic cameras, which detected malignant breast disorders with a sensitivity of 97% and a negative predictive value of 82%.

In particular, it is suggested that IRT may identify thermal abnormalities years before structural changes, which can be identified by ultrasound and mammography. For example, Gautherie and Gross [43] found that 38% of patients with abnormal infrared images were diagnosed with breast cancer in the four years following the abnormal infrared images.

In addition to passive imaging, cold stimulation-based imaging procedures are also practiced [44]. Blood vessels produced by cancerous tumors are simple endothelial tubes devoid of a muscular layer. Such blood vessels fail to constrict in response to sympathetic stimuli like sudden cold stress and show a hyperthermic pattern due to vasodilatation.

Deng and Liu [45] used fluid evaporation to enhance thermographic contrast in case of tumors underneath the skin.

8.2 THERMAL IMAGING: PRACTICAL CONSIDERATIONS

The utility of thermography in medicine is based on the early finding by Hardy [46] that the human skin, regardless of color, is a highly efficient radiator with an emissivity at 0.98, close to that of a perfect black body.

Interpretation of thermographic images requires experience. Also, thermographic imaging can be affected by multiple factors. Thus, in addition to the patient's acclimation, control of the ambient temperature, humidity, drafts, ventilation, direct sunlight, and other heat sources is required. Moreover, the use of standard protocols during thermographic imaging is critical. At least, to some degree, the inconsistencies between different studies can be attributed to differences between protocols.

In addition to the controlled environment and standardized protocols, several other important factors need to be considered for thermographic imaging and its interpretation.

8.2.1 Viewing Angle

An object's emissivity and apparent temperature vary with the viewing angle (see, for example, [1]). Thus, in practical applications, the viewing angle with respect to the tissue surface should not exceed 60° or, ideally, 45°, since for larger angles, the emissivity of the surface will be significantly reduced. Thus, for skin viewed obliquely, the lower emissivity may reduce the apparent temperature of more than 4°C, so a significant "hotspot" in this region might be undetected in a thermogram.

8.2.2 Core vs. Skin Temperature

Core body temperature refers to the temperature of the body's internal organs, such as the heart, liver, brain, and blood. The core body temperature is constant over time. The average normal body temperature is generally accepted as 37ºC.

Unlike the core body temperature, which is kept constant, the skin temperature is subjected to changes and is usually lower than the core.

In realistic conditions, the skin temperature is the balance between heat production and radiative, convective, and evaporative heat losses. In [47], it was found that other than the ambient temperature, blood perfusion and epidermis thickness are the primary factors responsible for skin temperature variations. In particular, the primary temperature drop in the skin is attributed to the cooling of the blood in the venous plexus. The temperature drop in the epidermis is on the scale of $0.1°C$ for the normal epidermis but can be $1.5-2°C$ or higher in calluses. Thus, local skin temperature variations can indicate epidermis thickness variations, particularly in callus-prone areas [48]. Free moisture on the skin (e.g., wet wound) significantly increases the heat transfer, resulting in a temperature drop of several degrees Celsius. The relative air humidity significantly contributes (by slowing heat dissipation) only in the case of evaporative heat loss from wet skin. Therefore, wet skin is undesirable and should be avoided during a thermographic assessment.

8.2.3 Wound Bed vs. Peri-Wound

As has been mentioned already, the presence of free moisture on the skin (e.g., wet wound) significantly increases the heat transfer, resulting in a skin temperature drop, which can be on the scale of several degrees Celsius. It leads to a well-known problem of thermographic image distortion caused by evaporative water loss in the wound bed. This problem can be solved by allowing the wound to dry completely (which may delay the timing of the assessment) or by applying a non-permeable covering to the wound bed, eliminating the evaporation problem [49].

8.2.4 Absolute vs. Relative Temperature

Because skin temperatures are highly variable between individuals, absolute temperatures have little value. Thus, interpreting an absolute skin temperature measurement (especially a single-point measurement) may pose a challenge because it is affected by multiple factors, both internal and external. Therefore, a temperature gradient (the difference in temperatures between two points) is a more objective measure, which will depend less on the ambient conditions.

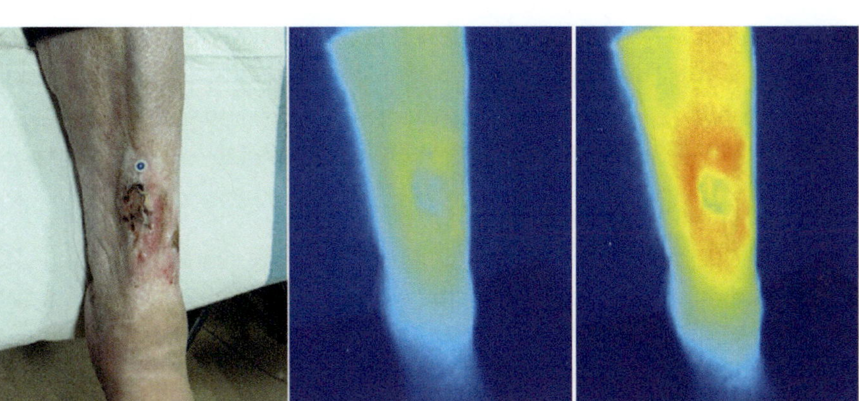

FIGURE 8.4 Use of absolute vs. relative temperature gradients for infrared thermal imaging. These panels show a traumatic ulcer in a diabetic individual. The middle panel reports the absolute temperature distribution in °C, while the right one is the temperature relative to the circle on the middle panel (reference point). A temperature gradient maximizes differences in the wound bed and peri-wound area while confounding factors are adjusted. (Photograph courtesy of Dr. Jose Ramirez-GarciaLuna).

8.2.5 Thermal Asymmetry

Relative temperature measurements have an important particular case, which can be termed "*thermal asymmetry*," where a region of interest (ROI) is compared to the contralateral region of the patient.

The thermal asymmetry concept is based on bilateral symmetry, one of the most important characteristics of the human body's surface heat pattern [50]. For example, Lahiri et al. [5] compiled the contralateral (left vs. right) temperature differences of various body segments in normal human subjects from several sources. They found that it does not exceed 0.5°C.

An illustration of the thermal asymmetry concept is the "*side-to-side*" temperature differences *between* feet are greater in patients with diabetes than among healthy controls, where there is little asymmetry between feet. Skin temperature usually shows symmetrical bilateral distribution; thus, *strong asymmetry* indicates an abnormality that can serve as a screening marker for more advanced tests. Skin temperature variations over 2.0°C may be helpful for identifying pathological situations in the diabetic foot [51].

For patients with diabetes (DM), mean temperatures significantly differed between the subgroups with angiopathy, neuropathy, and neuroischemia. Notably, neuropathic feet were warmest, followed by neuroischemia and then angiopathy without neuropathy. Patients with DM with neuropathy (27%) had a higher mean temperature compared to the healthy controls (plantar side: 29.4°C vs. 27.7°C ($p = 0.001$) [51].

8.2.6 Dynamic Thermography

Dynamic thermography (DT), which uses measurements before and after a physiologic challenge, is a useful clinical tool. For example, the six-minute walking test (6MWT) is an established physiologic challenge to evaluate PAD.

Huang et al. investigated the IRT measurements of the anterior shin and plantar foot before and after the 6MWT and the correlation with ABI measurements [52]. The exercise-induced temperature change (ETC) was defined as the post-exercise temperature minus the pre-exercise temperature. The resting temperatures were similar in PAD and non-PAD patients. However, the post-exercise temperature dropped in the lower extremities with arterial stenosis but was maintained or elevated slightly in the extremities with patent arteries. The temperature changes in PAD vs. non-PAD patients were -1.25°C vs -0.15°C $(p < 0.001)$. The ETC changes at the sole positively correlated with the 6MWT (Spearman correlation coefficient r $= 0.31$ $(p < 0.05)$, and it correlated with ABI measurements (Spearman correlation coefficient r $= 0.48$ $(p < 0.001)$.

Another approach to dynamic thermography is the fluid evaporation technique. For example, fluid evaporation can be used to detect cutaneous perforators in flaps, which can be performed by spraying the flap donor site with isopropyl alcohol followed by thermography to identify the precise location of flap perforators [53]. A similar approach can be used to enhance cancer visualization [45].

8.3 CONCLUSIONS

Thermography refers to passive noncontact spatially resolved temperature measurements over a large area. Noncontact temperature measurement techniques do not measure the temperature directly. Instead, they measure energy flow from the object and derive temperature from these measurements.

Thermal imaging sensors operate in the middle (MWIR, 3–5 µm) or long (LWIR, 8–14 µm) wavelength infrared (IR) spectral ranges. Thermal imaging sensors for biomedical research and clinical applications use low-cost, uncooled FPA microbolometers in the LWIR range.

Thermography in a medical context measures temperature distribution in superficial tissues. The skin temperature is affected by multiple factors, including environmental factors. However, in thermal equilibrium, it depends primarily on the blood supply. As a result, the primary clinical utility of thermography is to detect abnormalities in blood circulation.

Thermography has been studied extensively across different medical domains. Most widely, it is used for fever screening. Thermography has many applications in wound care, including infection and inflammation detection. In oncology, its primary utility is in breast cancer screening.

Given the strong dependency on numerous environmental factors, precautions must be taken during clinical thermographic imaging. Moreover, a temperature gradient (the difference in temperatures between two points) is more objective, as it is less dependent on the ambient conditions than absolute temperature measurements.

REFERENCES

1. JL Ramirez-GarciaLuna, R Bartlett, JE Arriaga-Caballero, et al. "Infrared thermography in wound care, surgery, and sports medicine: A review". *Front Physiol* 13: 838528 (2022).
2. J Martín-Vaquero, A Hernández Encinas, A Queiruga-Dios, et al. "Review on wearables to monitor foot temperature in diabetic patients". *Sensors (Basel)* 19(4): 776 (2019).
3. M Czerny. "Über photographie in ultraroten". *Z Physik* 53: 1–12 (1929).
4. E Schwamm, J Reeh. "Die Ultrarotstrahlung des Menschenund seine Molekularspektroskopie". *Hippokrates* 24: 737–742 (1953).
5. BB Lahiri, S Bagavathiappan, T Jayakumar, J Philip. "Medical applications of infrared thermography: A review". *Infrared Phys Technol* 55(4): 221–235 (2012).
6. R Gulias-Cañizo, ME Rodríguez-Malagón, L Botello-González, et al. "Applications of infrared thermography in ophthalmology". *Life (Basel)* 13(3): 723 (2023).
7. RN Lawson, GD Wlodek, DR Webster. "Thermographic assessment of burns and frostbite". *Canad Med Assos J* 84: 1129–31 (1961).
8. M Bharara, J Schoess, DG Armstrong. "Coming events cast their shadows before: Detecting inflammation in the acute diabetic foot and the foot in remission". *Diab Metab Res Rev* 28(Suppl. 1): 15–20 (2012).
9. L De Weerd, JB Mercer, S Weum. "Dynamic infrared termography". *Clin Plast Surg* 38: 277–292 (2011).
10. J Philip, T Jayakumar, B Raj, et al. "Infrared thermal imaging for detection of peripheral vascular disorders". *J Med Phys* 34: 43–47 (2009).
11. A Ilo, P Romsi, M Pokela, J Mäkelä. "Infrared thermography follow-up after lower limb revascularization". *J Diab Sci Technol* 15: 807–815 (2021).

12. GG Hallock. "Dynamic infrared thermography and smartphone thermal imaging as an adjunct for preoperative, intraoperative, and postoperative perforator free flap monitoring". *Plast Aesthet Res* 6: 29 (2019).

13. N Pereira, D Valenzuela, G Mangelsdorff, et al. "Detection of perforators for free flap planning using smartphone thermal imaging: A concordance study with computed tomographic angiography in 120 perforators". *Plast Reconstr Surg* 141: 787–792 (2018).

14. GG Hallock. "Doppler sonography and color duplex imaging for planning a perforator flap". *Clin Plast Surg* 30: 347–357 (2003).

15. D Chubb, WM Rozen, IS Whitaker, MW Ashton. "Digital thermographic photography ("thermal imaging") for preoperative perforator mapping". *Ann Plast Surg* 66: 324–325 (2011).

16. JM Cavaillon. "Once upon a time, inflammation". *J Venom Anim Toxins Incl Trop Dis* 27: e20200147 (2021).

17. S Tejiram, SP Tranchina, TE Travis, JW Shupp. "The first 24 hours: Burn shock resuscitation and early complications". *Surg Clin North Am* 103(3): 403–413 (2023).

18. JL Ramirez-GarciaLuna, K Rangel-Berridi, R Bartlett, et al. "Use of infrared thermal imaging for assessing acute inflammatory changes: A case series". *Cureus* 14(9): e28980 (2022).

19. M Schiavenato, RG Thiele. "Thermography detects subclinical inflammation in chronic tophaceous gout". *J Rheumatol* 39(1): 182–183 (2012).

20. JL Ramirez-GarciaLuna, SC Wang, T Yangzom, et al. "Use of thermal imaging and a dedicated wound-imaging smartphone app as an adjunct to staging hidradenitis suppurativa". *Br J Dermatol* 186(4): 723–726 (2022).

21. SJ Benbow, AW Chan, DR Bowsher, et al. "The prediction of diabetic neuropathic plantar foot ulceration by liquid-crystal contact thermography". *Diab Care* 17(8): 835–839 (1994).

22. N Kaabouch, WC Hu, Y Chen, et al. "Predicting neuropathic ulceration: Analysis of static temperature distributions in thermal images". *J of Biomed Opt* 15(6): 061715 (2010).

23. LA Lavery, KR Higgins, DR Lanctot, et al. "Home monitoring of foot skin temperatures to prevent ulceration". *Diab Care* 27: 2642–2647 (2004).

24. RG Frykberg, IL Gordon, AM Reyzelman, et al. "Feasibility and efficacy of a smart mat technology to predict development of diabetic plantar ulcers". *Diab Care* 40: 973–980 (2017).

25. T Nagase, H Sanada, K Takehara, et al. "Variations of plantar thermographic patterns in normal controls and non-ulcer diabetic patients: Novel classification using angiosome concept". *J Plast Reconst Aesth Surg JPRAS* 64(7): 860–866 (2011).

26. M Bharara, J Schoess, A Nouvong, DG Armstrong. "Wound inflammatory index: A 'proof of concept' study to assess wound healing trajectory". *J Diab Sci Technol* 4(4): 773–779 (2010).

27. LA Lavery, BJ Petersen, DR Linders, et al. "Unilateral remote temperature monitoring to predict future ulceration for the diabetic foot in remission". *BMJ Open Diab Res Care* 7: e000696 (2019).

28. TC Carpenter, S Schomberg, KR Stenmark. "Endothelin-mediated increases in lung VEGF content promote vascular leak in young rats exposed to viral infection and hypoxia". *Am J Physiol Lung Cell Mol Physiol* 289(6): L1075–L1082 (2005).

29. TE Serena, JR Hanft, R Snyder. "The lack of reliability of clinical examination in the diagnosis of wound infection: Preliminary communication". *Int J Low Extrem Wounds* 7(1): 32–35 (2008).

30. T Swanson, K Ousey, E Haesler, et al. "IWII wound infection in clinical practice consensus document: 2022 update". *J Wound Care* 31(Suppl. 12): S10–S21 (2022).

31. KY Woo, RG Sibbald. "A cross-sectional validation study of using NERDS and STONEES to assess bacterial burden". *Ostomy Wound Manage* 55(8): 40–48 (2009).

32. O Rahbek, HC Husum, M Fridberg, et al. "Intrarater reliability of digital thermography in detecting pin site infection: A proof of concept study". *Strategies Trauma Limb Reconstr* 16(1): 1–7 (2021).

33. A Chanmugam, D Langemo, K Thomason, et al. "Relative temperature maximum in wound infection and inflammation as compared with a control subject using long-wave infrared thermography". *Adv Skin Wound Care* 30(9): 406–414 (2017).

34. M Oe, RR Yotsu, H Sanada, et al. "Thermographic findings in a case of type 2 diabetes with foot ulcer and osteomyelitis". *J Wound Care* 21(6): 274–278 (2012).

35. K Lloyd Williams. "A thermographic prognostic index". In *Recent Advances in Medical Thermology*, EFJ Ring, B Phillips, Eds. Plenum Press, London, pp. 551–555 (1984).

36. P Gamagami. *Atlas of Mammography: New Early Signs in Breast Cancer.* Blackwell Science, United Kingdom (1986).

37. M Anbar. "Clinical thermal imaging today". *IEEE Eng Med Biol Mag* 17: 25–33 (1998).

38. A Di Carlo, F Elia, F Desiderio, et al. "Can video thermography improve differential diagnosis and therapy between basal cell carcinoma and actinic keratosis?" *Dermatol Ther* 27(5): 290–297 (2014).

39. R Lawson. "Implications of surface temperatures in the diagnosis of breast cancer". *Can Med Assoc J* 75: 309–311 (1956).

40. SA Feig, GS Shaber, GF Schwartz, et al. "Thermography, mammography, and clinical examination in breast cancer screening. Review of 16,000 studies". *Radiology* 122: 123–127 (1977).

41. D Kennedy, T Lee, D Seely. "A comparative review of thermography as a breast screening technique". *Integr Cancer Ther* 8: 9–16 (2009).

42. N Arora, D Martins, D Ruggerio, et al. "Effectiveness of a noninvasive digital infrared thermal imaging system in the detection of breast cancer". *Am J Surg* 196: 523–526 (2008).

43. M Gautherie, CM Gros. "Breast thermography and cancer risk prediction". *Cancer* 45: 51–56 (1980).

44. WC Amalu, WB Hobbins, JF Head, RL Elliot. "Infrared imaging of the breast – an overview". In *Biomedical Engineering Handbook, Medical Devices and Systems*, 3rd ed., JD Bronzino, Ed. CRC Press, p. 20 (2006).

45. ZS Deng, J Liu. "Enhancement of thermal diagnostics on tumors underneath the skin by induced evaporation". In *Proceedings of the 27th Annual Conference of IEEE Engineering in Medicine and Biology*, Shanghai, China (2005).

46. JD Hardy. "The radiation of heat from the human body". *J Clin Invest* 13: 539–615 (1934).

47. G Saiko. "Skin temperature: The impact of perfusion, epidermis thickness, and skin wetness". *Appl Sci* 12: 7106 (2022).

48. F Sadrzadeh-Afsharazar, R Raizman, G Saiko. "Utility of thermographic imaging for callus identification in wound and foot care". *Sensors* 23: 9376 (2023).

49. RP Cole, PG Shakespeare, HG Chissell, SG Jones. "Thermographic assessment of burns using a nonpermeable membrane as wound covering". *Burns* 17(2): 117–122 (1991).

50. R Vardasca. "Symmetry of temperature distribution in the upper and lower extremities". *Thermol Int* 18: 154–15 (2008).

51. A Ilo, P Romsi, J Mäkelä. "Infrared thermography and vascular disorders in diabetic feet". *J Diabetes Sci Technol* 14(1): 28–36 (2020).

52. CL Huang, YW Wu, CL Hwang, et al. "The application of infrared thermography in evaluation of patients at high risk for lower extremity peripheral arterial disease". *J Vasc Surg* 54(4): 1074–1080 (2011).

53. MV Muntean, S Strilciuc, F Ardelean, AV Georgescu. "Dynamic infrared mapping of cutaneous perforators". *J Xiangya Med* 3: 16 (2018).

CHAPTER 9

Vibrational and Terahertz Techniques in Physiological Optical Imaging

THE VIBRATIONAL AND TERAHERTZ (THz) spectroscopy and imaging aim to probe collective motions of biomolecules, such as vibration, rotation, and libration. These characteristic energies make these spectroscopy/imaging techniques an excellent tool for biological and medical studies, exhibiting high sensitivity in measurements targeting changes in biomolecules, cells, and tissues. With advances in sensors and sources, this group of technologies is rapidly evolving, including transitioning from traditional spectroscopic into imaging modes.

Vibrational imaging is based on vibrational spectroscopy. As the vibrational states of a molecule in the ground electronic state can be probed either by directly measuring the absolute frequency (IR absorption) or the relative frequency or Raman shift (Stokes and anti-Stokes) of the allowed transitions (see Figure 9.1), vibrational spectroscopy can be split into two large fields of studies: IR spectroscopy and Raman spectroscopy.

While both Raman and IR spectroscopy share the same band range[1] up to 4,000 cm^{-1}, they have very different mechanisms and characteristics. IR

DOI: 10.1201/9781003482505-11

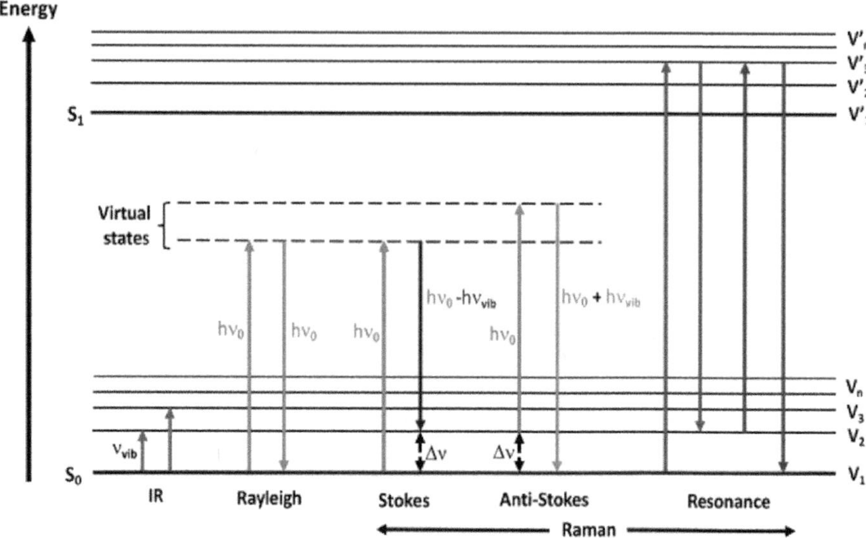

FIGURE 9.1 Jablonski energy diagram showing the transitions involved during infrared absorption, Rayleigh, Raman Stokes, anti-Stokes, and Resonance Raman scattering. The vibrational states $\left(V_n\right)$ of a molecule in the ground electronic state $\left(S_0\right)$ can be probed either by directly measuring the absolute frequency (IR absorption) or the relative frequency or Raman shift (Stokes and anti-Stokes) of the allowed transitions. Resonance Raman also involves the vibrational states $\left(V_n'\right)$ of the excited electronic state $\left(S_1\right)$. $h\nu_0$ = incident laser energy, $h\nu_{vib}$ = vibrational energy, $\Delta\nu$ = Raman shift and ν_{vib} = vibrational frequencies. Reproduced from [1] under CC BY 4.0 license.

absorption results from changes in the dipole moment. As a result, asymmetric and polar molecules show stronger IR spectra compared to more symmetric nonpolar molecules. Consequently, most organic molecules show IR spectra with many overlapping bands. Unlike IR, Raman activity depends on changes in the polarizability tensor caused by molecular bond vibrations. As a result, only symmetric modes and nonpolar oscillating molecules give strong Raman spectra [1]. As such, IR and Raman's spectroscopies provide different information about biofluids and biotissues. Thus, we will consider these two approaches separately.

Vibrational spectroscopy provides valuable information about analytes. For biological samples, approximately 90% of the peaks are found in the spectral region, covering $\Delta\nu \sim 500\ \mathrm{cm^{-1}}$ to $1800\ \mathrm{cm^{-1}}$, with the remaining found in the higher energy CH/OH stretching vibrational modes covering $\Delta\tilde{v} \sim 2700$ cm^{-1} to 3300 cm^{-1} [2].

This combination of peaks is quite unique ("fingerprint") and can be used to identify molecules with high specificity. Thus, the 500–1800 cm^{-1} region is often called the fingerprint region because it provides a unique pattern or "fingerprint" of the molecule.

Currently, vibrational and terahertz imaging have mostly preclinical *in vitro* applications in body fluids and cells, *ex vivo* tissues, and *in vivo* small animals. We will discuss briefly their clinical translation.

9.1 RAMAN IMAGING

While elastic scattering is the predominant mechanism of electromagnetic wave scattering in turbid media, some photons may exchange energy with a scatterer (a molecule in this case) and excite the molecule's vibration/rotational vibration states. In this type of scattering, called inelastic or Raman scattering, the energy is transferred between the photon and the scattering center (see Fig.9.1). As a result, the scattered photons usually have lower energy than the incident photons (Stokes shift). However, in certain circumstances, the scattered photons may have higher energy than the incident photons (anti-Stokes shift). By convention, the change in the photon energy in Raman spectroscopy (Stock shift) is characterized by a wave number shift, $\Delta\tilde{\nu}$, measured in [cm^{-1}]. The lifetime of Raman scattering, τ_R, is less than 10^{-14} s.

Inelastic scattering has a resonant origin, with peaks corresponding to the transitions between vibration/rotational-vibration states of a particular molecule (see Figure 9.1). In general, inelastic scattering is a rare event that happens at the rate of 10^{-8} of the elastic scattering [3]. In addition, the Stock shift for Raman scattering is relatively small. Thus, spontaneous Raman scattering can generally be observed in lab conditions with laser sources and specialized optics. However, multiple techniques can increase the Raman signal strength, including surface-enhanced Raman scattering (SERS) [4], tip-enhanced Raman scattering (TERS) [5], stimulated Raman scattering (SRS) [6], and coherent anti-Stokes Raman scattering (CARS) [7]. Raman enhancement techniques like SERS allow significant (on a scale of several orders of magnitude) increases in signal strength and SNR, thus increasing the method's sensitivity. Combined with very high specificity, Raman spectroscopy is a unique tool for many applications in biomedicine and beyond.

It should be pointed out that unlike fluorescence, which is sensitive to excitation light wavelength in a resonant manner (although usually, they are pretty broad peaks), the Raman scattering can be excited by light in a wide range of wavelengths as its signal intensity is inversely proportional to the wavelength of the excitation light.

9.1.1 Clinical Utility

Recent developments in Raman imaging for biomedical applications were reviewed by Krafft et al. [8]. However, Raman imaging is still in its infancy, and most clinical studies use Raman spectroscopy. Thus, we will discuss the clinical utility of Raman spectroscopy.

9.1.1.1 In Vivo

The progress in clinical translation of Raman spectroscopy for cancer detection and surgical guidance was reviewed by Santos et al. [9]. *In vivo* Raman spectroscopy/imaging applications can be roughly split into diagnostics and intraoperative guidance.

9.1.1.1.1 Diagnostics   Skin. Lieber et al. [10] used a Raman fiber probe to measure lesions from non-melanoma skin cancers and demonstrated 100% sensitivity and 91% specificity in discriminating these lesions from normal tissues in a small study. In a study with over 1,000 cases, Lui et al. [11], using a handheld Raman probe measuring spectra in less than 1 s, were able to identify malignant melanoma lesions with high sensitivity >90% but low specificity, ranging between 15% and 54%, when discriminating malignant melanoma from non-melanoma pigmented lesions and seborrheic keratosis.

Endoscopy. The capacity of Raman spectroscopy to discriminate between normal and malignant tissues *in vivo* was demonstrated for various hollow organs using endoscopy. For example, using a fiber probe with 5 s acquisition times in Barrett's esophagus, the detection of HGD adenocarcinoma was achieved with high sensitivity (86%) and specificity (98%) [12]. Lin et al.[13] demonstrated *in vivo* diagnosis of laryngeal carcinomas. Short et al. [14] used Raman spectroscopy to increase the specificity of cancer detection by autofluorescence during bronchoscopy. They achieved a sensitivity of 96% and a specificity of 91% for discrimination of preneoplastic lesions.

9.1.1.1.2 Intraoperative Analysis   The primary utility of Raman measurements for intraoperative analysis is to provide the surgeon with a real-time measure of tumor margin analysis or metastatic lesions such as sentinel lymph nodes [15]. Multiple teams have demonstrated this utility for different types of cancers.

For example, Lloyd et al. [16] were able to separate head and neck lymphoma tissues into reactive nodes (swollen from reaction to infection), primary malignancies (lymphomas), and secondary malignancies (metastatic squamous cell carcinomas and adenocarcinomas) with 90% sensitivity and 86% specificity.

The close to real-time surgical guidance is particularly important for Mohs surgery, where the tissue is removed layer by layer until no tumor is found on histopathological assessment. The use of wide-field autofluorescence imaging of skin tissues to enable rapid identification of regions of concern for localized Raman measurements during Mohs surgery shows great promise for enabling dermatological surgeons to obtain an accurate measure of basal cell carcinoma margins in the operating theatre, rather than waiting for pathological analysis [17].

9.1.1.2 Ex Vivo

9.1.1.2.1 Margin Detection   In addition to being used for margin detection *in vivo*, Raman spectroscopy and imaging can be applied to freshly excised tissue samples to provide real-time surgical guidance. The significant advantage of the Raman spectroscopy is its nondestructive nature. As such, the samples after Raman measurements can go through routine histological staining.

9.1.1.2.2 Detection of Pathogens in Biofluids   Infectious diseases can be diagnosed in biofluids with Raman using several strategies [15]: a) analysis of complete sample composition, b) detection of defined DNA/RNA sequences, and c) specific detection of antigens on the cell surface.

However, analyzing highly complex samples like body fluids using Raman spectroscopy entails certain challenges. Thus, complex sample preparation strategies (including at least centrifugation) and complex data processing need to be applied depending on the assay.

Urine. Bacterial cell concentrations of urine samples from patients with urinary tract infection (UTI) can range from 10^2 to 10^5 cells/ml. Thus, typically, a sample preparation to achieve higher concentrations is required. Premasiri et al. [18] used SERS with Au nanoparticles to investigate UTI-relevant bacteria after several centrifugation and washing steps.

Blood. Detecting pathogens directly in the whole blood matrix, which contains billions of blood cells, is exceptionally challenging since, quite often, there are only 10 CFU/ml or fewer bacteria present. Thus, several strategies were suggested. Boardman et al. [19] developed a preprocessing routine (selective lysis plus centrifugation) for whole blood that concentrates viable microorganisms from a 10 ml sample to a 200 μl volume. Ngo et al. [20] developed a SERS assay for analyzing whole blood samples to identify the malaria parasite *Plasmodium falciparum*. Aiming at a nucleic-based identification, they achieved a detection limit of 200 fM with synthetic target DNA.

In addition to the whole blood, blood serum and plasma are also frequently used in routine clinical diagnostics. Neugebauer et al. [21] analyzed blood plasma from ICU patients either with systemic inflammatory response syndrome (SIRS) or sepsis using Raman spectroscopy. They were able to distinguish between SIRS and sepsis with 80% accuracy.

Other biofluids. Ghebremedhin et al. [22] used SERS to detect bacterial presence in wound exudates. Using resonance Raman spectroscopy, Gonchukov et al. [23] detected carotenoids as biomarkers for periodontitis in dried saliva samples from periodontitis patients.

9.1.2 Practical Considerations

Despite various developed enhancement techniques, most studies use traditional Raman technology. As such, autofluorescence of the tissue can significantly impact the weak Raman signal. Moreover, in the most typical scenario, the fiber-optic probe is used, which also exhibits autofluorescence (for example, the autofluorescence in flat and tapered optical fibers was studied by Bianco et al. [24]).

Thus, certain strategies to tackle these artifacts need to be employed. Strategies to optimize probes for *in vivo* Raman spectroscopy were reviewed by Stevens et al. [25]. Another approach is to optimize the light source wavelength. While shorter wavelengths (blue visible range and UV) generate a stronger signal, they also excite autofluorescence of the tissue, which may mask the weak Raman signal. Consequently, the longer range of visible and NIR (500–830 nm) is typically used, with a 785 nm laser diode as the most common source, which balances the competing factors between Raman signal intensity, fluorescence, detector sensitivity, and cost. However, for samples with strong autofluorescence, such as dyes, a 1,064 nm laser may be needed. For other applications, lasers in blue and green ranges (e.g., 532 nm) are becoming more common in Raman spectroscopy.

Soliman et al. [26] compared the performance of eight commercial Raman spectrometers ranging in size from benchtop Raman microscopes to portable and handheld Raman spectrometers to detect two well-established biomarkers for evaluating cardiovascular health (cardiac troponin I and heart fatty acid binding protein) and assessed their possibility to be used in POC settings. They found that the 780 and 785 nm laser sources exhibited a reduced background signal and provided higher sensitivity than those with 633 and 638 nm laser sources. Furthermore, the spectrometer equipped with the single acquisition line readout functionality improves performance compared to the point scan spectrometers and makes measurements faster and easier.

9.2 IR IMAGING

Infrared spectroscopy can be defined as the study of the absorption properties of materials arising from changes in their molecular vibrational motions upon interaction with an IR source.

The primary approach in IR imaging is the coupling of Fourier-transform infrared (FTIR) spectrometers to optical microscopes, which is often called FTIR micro-spectroscopy (FTIRM). As a result, FTIRM provides spatially resolved IR spectra.

It should be noted that IR is not transmitted by glass. Consequently, all the instrumentation optics, including the microscope objectives, are mirror based. Mid-IR mirrors have recently achieved impressive quality (so-called supermirrors). In particular, 400,000 finesse has been achieved [27].

Typically, the FTIR setup is based on an interferometer, which converts the observed time-dependent signal (interferogram) into the frequency domain using a Fourier transform.

To obtain spatially resolved 2D IR spectra, two optical configurations are used: a) mapping or scanning techniques, and b) imaging techniques with focal plane array (FPA) detectors.

In the FTIR mapping approach, IR spectra are acquired sequentially point by point over the area of interest using a single-element detector and a motorized sample stage. The amount of the radiation arriving at the sample plane is regulated by an aperture, which determines the lateral resolution of the mapping (scanning) technique. A high lateral resolution near the diffraction limit of IR radiation can be achieved with small apertures $\left(\leq 10 \times 10\ \mu\text{m}^2\right)$, which dramatically restricts the IR flux that reaches the detector. Thus, for traditional thermal sources, it results in long acquisition times or low SNR.

9.2.1 Clinical Utility

FTIR spectra, particularly in the 500–1800 cm⁻¹ (fingerprint) region, provide valuable information about analytes.

In medicine, IR spectroscopy/imaging can be used to screen a field of cells from a blood smear or tissue section based on the cell's intrinsic molecular phenotype, which represents a potentially significant step forward compared to conventional hematoxylin and eosin (H&E) and immune-chemical staining methods.

As samples require minimal or no preparation, IR measurements can be categorized as *ex vivo*.

According to Pahlow et al. [15], the utility of IR imaging and spectroscopy in POC settings is in three areas: a) imaging of biofluids, b) single cell analysis, and c) diagnosis of tissue sections.

Spectroscopic techniques are well suited to the analysis of biofluids. Infrared bands can be related to different constituents of the metabolome, proteome, and lipidome of biofluids such as urine, [28] whole blood, [29] saliva, [30] or cerebrospinal fluid [31]. Infrared spectra can be acquired in a few minutes without complex sample preparation, making this technique suitable for point-of-care (POC) analysis.

Despite limitations on spatial resolution, FTIRM can be applied for single-cell analysis. In particular, as the different stages of the *Plasmodium* (malaria parasite) life cycle could be discriminated by their IR absorbance in the CH-stretching region (3,100–2,800 cm⁻¹) [32], FTIRM was used for the detection of red blood cells infected with *P. falciparum* in the trophozoite stage [33].

FTIRM has been applied in a diagnostic context for various tissues based on the biochemical fingerprint of the cells that constitute the tissue section. At the current stage, the technology was applied only to biopsies. Bone quality was, for example, assessed by FTIRM and Raman imaging by Kimura-Suda et al. [34] by observing the PO_4^{3-} band between $1200-900\ cm^{-1}$, which is affected by changes in the calcium phosphate composition and indicates bone maturity. In oncology, FTIRM can be used to differentiate between normal and cancerous tissues. In particular, FTIRM was used for the diagnosis of breast cancer [35].

9.2.2 Practical Considerations

Despite its advantages, such as molecular specificity, FTIR spectroscopy suffers from some shortcomings that severely limit its application to biological samples – mainly sensitivity issues and difficulties in studying aqueous solutions.

The low sensitivity can be overcome using plasmonic resonance amplification in metallic nanoparticles [36], which is called surface-enhanced infrared absorption (SEIRA) [37]. It allows the *in situ* monitoring of proteins and nanoparticle interactions in aqueous media at high sensitivity and in real time [38].

The absorption of water in the mid-IR region is very intense, as its OH-bending absorption is much stronger than any signal from the protein samples. Thus, some strategies (like dehydration) need to be employed to analyze biological samples. However, it limits its clinical applications.

Additionally, despite advances in detector technologies, there are still issues with mid-IR sources. Currently, there are three types of IR sources. Traditional thermal sources are characterized by low flux and long acquisition times. Synchrotron radiation sources, which are 100–1000x brighter than thermal sources, have an obvious shortcoming due to their availability. Semiconductor quantum-cascade lasers (QCLs) provide the best brightness (up to three orders of magnitude higher than a synchrotron at certain wavelengths). However, even tunable QSRs have a pretty narrow band. A single-tuner laser covers 2–3 μm bandwidth. Thus, up to 4 QCL modules are required to cover the whole IR range.

With long acquisition times or complex setups, the technology is not well suited for clinical use. However, several technological developments may drive clinical translation.

For example, if the relevant spectral wavelengths are known, a simpler imaging approach to acquire only these data would enable data collection 100x faster than the traditional FTIR approach. For example, multiple studies show that only a few wavelengths over the broad mid-IR range are required to provide all the information required for tissue histopathology. Thus, QCLs enabled discrete frequencies IR (DFIR) technology to provide high throughput analysis in clinically relevant timescales [39].

Another approach that can be used to analyze aqueous biological samples is to use the attenuated total reflectance (ATR) technique [40]. In this technique, an IR beam is focused at a set angle onto a crystal with a high refractive index. It produces an evanescent standing wave resulting from internal reflections when it propagates through it. This wave reaches beyond the outer surface of the crystal and then into the sample by a few microns $(0.5 - 5\,\mu m)$, which is held in contact with it. However, ATR is an inheritably single-point methodology. It will be hard to translate it to an imaging modality. Moreover, the technique requires contact between the ATR crystal and the sample.

9.3 TERAHERTZ IMAGING

The terahertz (THz) range lies within the transition region between microwave and far infrared. It typically refers to electromagnetic waves between 0.1 and 10 THz, corresponding to 3 mm to 30 μm wavelengths. However, the exact boundaries can vary depending on the source. For example, the International Telecommunication Union (ITU) designates the band of frequencies from 0.3 to 3 terahertz (THz).

The THz energy spectrum covers the characteristic energies of biomolecular collective motions such as vibration, rotation, and libration, as well as the low energy of hydrogen bonds, which are abundant in water-based biomedical samples. These characteristic energies make THz spectroscopy/imaging an excellent source for biological and medical studies, exhibiting high sensitivity in measurements targeting changes in biomolecules, cells, and tissues.

Similarly to mid-IR, the THz radiation cannot reach deep into wet biological tissues because water molecules heavily attenuate it. In particular, the radiation absorbance of water molecules is greater than $200\,cm^{-1}$ at 1 THz and room temperature. As such, THz waves can penetrate only a couple of hundred micrometers into the human skin [41].

9.3.1 Technology

Terahertz technology has experienced significant advancements in recent years.

9.3.1.1 THz Radiation Sources

THz range radiation sources have been reviewed by Gallerano and Bierdon [42]. Aside from free electron lasers (FEL) and synchrotron THz sources, THz range radiation sources include multiple approaches suitable for the biomedical field, including other free electron-based systems (e.g., klystrons and gyrotrons), as well as solid-state oscillators (e.g., Gunn diodes) with frequency multipliers, gas lasers (far infrared gas lasers, e.g., methanol pumped by CO_2 laser), quantum cascade lasers, and laser-driven THz emitters (e.g., Ti: sapphire lasers). Solid-state oscillators and free electron-based systems typically cover the low-frequency side (from millimeter wave range). At the same time, gas and quantum cascade lasers come from the high-frequency side (far infrared range).

9.3.1.2 Detectors

THz detectors can be divided into two groups [43]:

- Incoherent detection systems (with direct detection sensors) allow only signal amplitude detection. As a rule, they are broadband detection systems and

- coherent detection systems, which allow the detection of not only the amplitude of the signal but also its phase. As a rule, they are narrowband systems.

Coherent signal detection systems use heterodyne circuit design since, so far, proper amplifiers do not exist for high radiation frequency ranges. The detected signals are transferred to much lower frequencies (~1–30 GHz), where low noise amplifiers amplify them.

Direct detection of THz radiation includes a wide range of devices, including Golay cells. However, not all of them are suitable for miniaturization. In particular, some of them have gas-filled cavities (Golay cells) or require cryogenic temperatures. Among potentially interesting uncooled approaches are microbolometers, [44] pyroelectric detectors,[45] Schottky diodes, [46] Si MOSFET, [47] Si FET, [48] and HgCdTe heterostructures [49]. Different types of THz sensors were reviewed by Sizov and Rogalski [43].

The noise equivalent power (NEP) for uncooled detectors typically is from 10^{-10} to 10^{-9} W/Hz$^{1/2}$.

The photoconductive detection of broadband THz radiation is based on antenna structures similar to those used to generate pulsed broadband THz radiation emission spectra. It is used in time domain spectroscopy (TDS) experiments or imaging. The detectors as photoconductive antennas (PCAs) on the base of highly resistive semiconductors (e.g., low-temperature grown GaAs or narrow-gap InGaAs) are frequently used in the TDS technique.

9.3.1.3 Systems

Two primary types of THz setups are used in the biomedical field: time-domain spectroscopy (TDS) and continuous waves (CW).

TDS is based on ultrafast lasers and employs a time-of-flight approach to measure the propagated pulses' amplitude reduction and time delay. From this information, the absorption and refractive index of the sample can be extracted.

General THz-TDS systems are based on laser-driven solid-state emitters, such as an 800-nm Ti sapphire oscillator or a 1.5-µm telecommunication fiber laser. In these devices, a short (femtosecond) laser beam pulse generates and accelerates carriers in the gap between closely spaced electrodes. As a result, the current transient radiates in a broad band at THz frequencies corresponding to the Fourier transform of the laser pulse time profile [43].

The typical frequency range covered by laser-driven solid-state emitters is 0.2 to 2 THz or higher, depending on the laser pulse parameters. Average power levels range from nanowatts to hundred microwatts, and pulse energies typically range from femtojoules to nanojoules [43].

The development of an uncooled bolometer array and THz quantum cascade lasers (QCL) enables real-time THz imaging [50].

9.3.2 Clinical Utility

Potential clinical applications of THz spectroscopy and imaging were summarized by Son et al. [51].

9.3.2.1 Oncology

Terahertz waves have several features that make them particularly suitable for cancer detection. Firstly, terahertz waves are highly sensitive to the presence of water due to the strong absorption of hydrogen bonds in the THz region. This property can be used to detect cancerous areas where the water content is higher than in normal tissue because of their excessive metabolism [52]. Secondly, Cheon et al. [53] observed a resonant feature at 1.65 THz associated with various types of cancer. This resonance is believed to originate from methylated DNA and might serve as a universal cancer biomarker.

9.3.2.1.1 *In Vivo*　Wallace et al. [54] reported THz imaging of skin cancer tissues *ex vivo* and *in vivo*. They used the terahertz pulsed imaging (TPI) system to differentiate basal cell carcinoma (BCC) from normal tissues in 18 *ex vivo* and five *in vivo* patients. They confirmed THz's potential to delineate tumor margins, as the diseased tissue showed a change in terahertz properties compared with normal tissue, manifested through a broadening of the reflected terahertz pulse. Regions of disease identified in the terahertz image correlated well with histology.

9.3.2.1.2 *Ex Vivo*　Gastric cancer. Ji et al. [55] reported that THz imaging was able to detect the tumor boundary using early gastric cancer (EGC) tissues sampled from eight patients. They identified four THz parameters that could be suggested for quantitative discrimination of EGCs.

Breast cancer. Fitzgerald et al. [56] used TPI to map the margins of exposed breast tumors. They investigated 22 excised human breast tissue specimens with carcinoma excised from 22 women and found that TPI can depict both invasive breast carcinoma and ductal carcinoma *in situ* under controlled conditions. Grootendorst et al. [57] used a handheld TPI instrument to discriminate benign from malignant breast tissue *ex vivo*. They analyzed 46 freshly excised breast cancer samples and compared their finding with histology. Using a support vector machine (SVM), they found that TPI achieved 75%, 86%, and 66% accuracy, sensitivity, and specificity, respectively.

Brain tumors. The brain is characterized by a high-lipid content due to the myelin sheath around neurons. As such, the myelin sheath, which contains both fat and proteins, provides a high contrast in THz images of brain cancer because the tumor contains elevated protein concentrations responsible for high THz radiation absorption, whereas lipids are less radiation absorbing. Ji et al. [59] used terahertz reflectometry imaging (TRI) to visualize tumor margins in grade II (low grade) vs. grade III and IV (high grade) gliomas on tissues from 14 glioma patients. They demonstrate that TRI provides tumor discrimination and delineation of tumor margins in brain tissues with high sensitivity (compared versus H&E stained image).

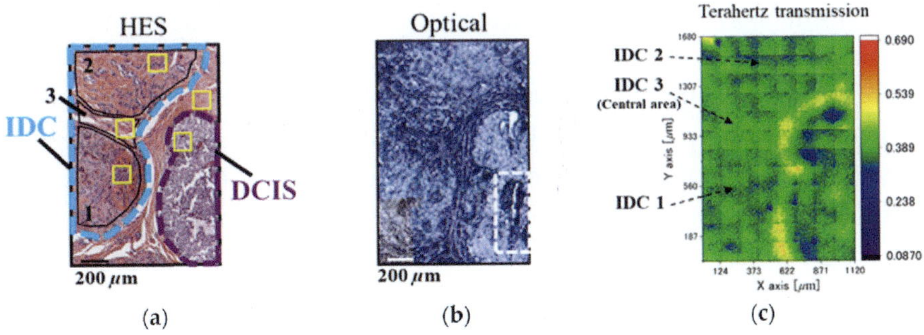

FIGURE 9.2   (a) Hematoxylin-eosin-saffron (HES), (b) Optical, and (c) Terahertz-transmission, images of a formalin-fixed paraffin-embedded (FFPE) human breast tissue section deposited on a GaAs (110) crystal. (Reproduced from [58] under CC BY 4.0 license).

9.3.2.2 In Vitro Diagnostics

THz devices and systems have a potential to rapidly differentiate tiny amounts of specimens, such as nucleic acids, proteins, and various metabolites, at the molecular level. As such, THz systems can be used in the development of in vitro diagnostic (IVD) medical devices, enabling early disease detection from sample specimens such as blood, body fluid, or breath.

George et al. [60] reported a THz microfluidic device that, from a small sample of 10 pmol, measures the absorbance spectrum of bovine serum albumin in the frequency range 0.5–2.5 THz.

THz systems were tried for glucose sensing. In particular, Chen et al. [61] reported a linear relationship between the blood glucose level and THz absorbance in 70 diabetic patients.

Analysis of gases originating from exhalation and flatus could provide information for diagnostics of multiple diseases like bronchial asthma and lung or colon cancers. THz spectroscopy could be employed as a high-sensitivity gas sensor. For example, Fosnight et al. [62] measured ethanol, methanol, and acetone gases at femtomole concentrations, which can be used to diagnose respiratory diseases like asthma or pemphigus.

9.3.3 Practical Considerations

The critical differences between detection at sub-mm wavelengths and IR detection lie in large photon wavelengths (e.g., $\lambda = 300$ µm). Thus, the diffraction limit determines the low spatial resolution of THz systems. Two approaches could be applied to achieve higher spatial imaging resolution: solid immersion lenses (usually Si lenses) and near-field imaging. However, THz near field has not been developed yet due to the lack of materials [43].

Terahertz imaging (or at least spectroscopy) is already suitable for endoscopic applications. For example, Ji et al. [63] fabricated and characterized a miniaturized optical fiber-coupled terahertz (THz) endoscope system with (2 x 4 mm) x 6 mm cross-section. The endoscopic system utilized a photoconductive generator and detector driven by a mode-locked Ti sapphire laser. The same group also designed and fabricated a THz otoscope with THz and conventional optical diagnostics [64].

Woodward et al. [65] proposed terahertz pulse imaging (TPI) and achieved resolutions at 1 THz of 350 µm laterally and 40 µm axially in the skin. As TPI is a coherent, time-gated, low-noise technique, both phase and amplitude information can be obtained, from which a medium's absorption and refractive index can be determined.

Ex vivo imaging can be performed using a scanning setup. For example, Ji et al. [55] used a movable sample stage to image a 50 × 47.5-mm area with a 500-μm scanning resolution. The total acquisition time per THz image was approximately 8 min.

As mentioned, THz radiation is characterized by a small penetration depth in biological tissues. Several techniques have been proposed to overcome this limitation. Firstly, freezing techniques were proposed to increase the penetration depth of THz radiation into wet tissues [66], as the absorption coefficient of ice is one order of magnitude lower than that of liquid water [52]. However, this approach is obviously limited in clinical practice. Secondly, the topical application of penetration-enhancing agents (PEA), like glycerol, may also increase penetration depth. Oh et al. [67] found that the intensity of the penetrated signal through fresh tissues when applying glycerol was two times larger than the regular THz signal. As glycerol is also an optical clearing agent (OCA), it is most likely that the enhancement can be linked to local dehydration (see section 4.3.2 for further details).

9.4 CONCLUSIONS

The vibrational and terahertz (THz) spectroscopy and imaging aim to probe collective motions of biomolecules, such as vibration, rotation, and libration. These high-specificity, label-free imaging modalities may significantly improve patient outcomes.

As water absorption in IR and terahertz ranges is very significant, the vibrational and terahertz (THz) spectroscopy and imaging are characterized by small penetration depths.

Vibrational and THz imaging/spectroscopy have not yet been translated to the clinic. At least partially, this can be attributed to the fact that technologies have just reached the level where biomedical applications have become a viable reality.

NOTE

1 By convention in vibrational spectroscopy, the wavenumbers (measured in [cm^{-1}]) are used instead of wavelengths. Wavelength can be converted in wavenumber using v [cm^{-1}] = 10,000/λ [μm].

REFERENCES

1. CFGC Geraldes. "Introduction to infrared and Raman-based biomedical molecular imaging and comparison with other modalities". *Molecules* 25(23): 5547 (2020).

2. CH Camp Jr, MT Cicerone. "Chemically sensitive bioimaging with coherent Raman scattering". *Nat Photonics* 9: 295–305 (2015).

3. HJ Bowley, DJ Gardiner, DL Gerrard, et al. *Practical Raman Spectroscopy*. Springer, Berlin, Heidelberg (1989).

4. M Fleischmann, PJ Hendra, AJ McQuillan. "Raman spectra of pyridine adsorbed at a silver electrode". *Chem Phys Lett* 26: 163–166 (1974).

5. MD Sonntag, JM Klingsporn, LK Garibay, et al. "Single-molecule tip-enhanced Raman spectroscopy". *J Phys Chem C* 116: 478–483 (2011).

6. MBJ Roeffaers, X Zhang, CW Freudiger, et al. "Label-free imaging of biomolecules in food products using stimulated Raman microscopy". *J Biomed Opt* 16: 021118 (2011).

7. J-X Cheng, XS Xie. "Coherent anti-stokes Raman scattering microscopy: Instrumentation, theory, and applications". *J Phys Chem B* 108: 827–840 (2004).

8. C Krafft, IW Schie, T Meyer, et al. "Developments in spontaneous and coherent Raman scattering microscopic imaging for biomedical applications". *Chem Soc Rev* 45(7): 1819–1849 (2016).

9. IP Santos, EM Barroso, TC Bakker Schut, et al. "Raman spectroscopy for cancer detection and cancer surgery guidance: Translation to the clinics". *Analyst* 142(17): 3025–3047 (2017).

10. CA Lieber, SK Majumder, DL Ellis, et al. "In vivo nonmelanoma skin cancer diagnosis using Raman microspectroscopy". *Lasers Surg Med* 40(7): 461–467 (2008).

11. H Lui, J Zhao, DI Mclean, H Zeng. "Real-time Raman spectroscopy for in vivo skin cancer diagnosis". *Cancer Res* 72(10): 2491–2500 (2012).

12. LM Almond, J Hutchings, G Lloyd, et al. "Endoscopic Raman spectroscopy enables objective diagnosis of dysplasia in Barrett's esophagus". *Gastrointest Endosc* 79(1): 37–45 (2014).

13. K Lin, W Zheng, CM Lim, Z Huang. "Real-time in vivo diagnosis of laryngeal carcinoma with rapid fiber-optic Raman spectroscopy". *Biomed Opt Exp* 7(9): 3705–3715 (2016).

14. MA Short, S Lam, AM McWilliams, et al. "Using laser Raman spectroscopy to reduce false positives of autofluorescence bronchoscopies: A pilot study". *J Thoracic Oncol* 6(7): 1206–1214 (2011).

15. S Pahlow, K Weber, J Popp, et al. "Application of vibrational spectroscopy and imaging to point-of-care medicine: A review". *Appl Spectrosc* 72: 52–84 (2018).

16. GR Lloyd, LE Orr, J Christie-Brown, et al. "Discrimination between benign, primary and secondary malignancies in lymph nodes from the head and neck utilising Raman spectroscopy and multivariate analysis". *Analyst* 138(14): 3900–3908 (2013).

17. K Kong, CJ Rowlands, S Varma, et al. "Diagnosis of tumors during tissue-conserving surgery with integrated autofluorescence and Raman scattering microscopy". *PNAS* 110(38): 15189–15194 (2013).

18. WR Premasiri, AF Sauer-Budge, JC Lee, et al. "Rapid bacterial diagnostics via surface enhanced Raman microscopy". *Spectroscopy* 27(6): S8–S31 (2012).

19. AK Boardman, WS Wong, WR Premasiri, et al. "Rapid detection of bacteria from blood with surface-enhanced Raman spectroscopy". *Anal Chem* 88(16): 8026–8035 (2016).

20. HT Ngo, E Freedman, RA Odion, et al. "Direct detection of unamplified pathogen RNA in blood lysate using an integrated lab-in-a-stick device and ultrabright SERS nanorattles". *Sci Rep* 8(1): 4075 (2018).

21. U Neugebauer, S Trenkmann, T Bocklitz, et al. "Fast differentiation of SIRS and sepsis from blood plasma of ICU patients using Raman spectroscopy". *J Biophotonics* 7(3–4): 232–240 (2014).

22. M Ghebremedhin, S Yesupriya, NJ Crane. "Surface enhanced Raman spectroscopy as a point-of-care diagnostic for infection in wound effluent". In *Proceedings of the Optical Diagnostics and Sensing XVI: Toward Point-of-Care Diagnostics*, p. 9715 (2016).

23. S Gonchukov, A Sukhinina, D Bakhmutov, et al. "Periodontitis diagnostics using resonance Raman spectroscopy on saliva". *Laser Phys Lett* 10(7): 075610 (2013).

24. M Bianco, A Balena, M Pisanello, et al. "Comparative study of autofluorescence in flat and tapered optical fibers towards application in depth-resolved fluorescence lifetime photometry in brain tissue". *Biomed Opt Express* 12(2): 993–1010 (2021).

25. O Stevens, IEI Petterson, JC Day, N Stone. "Developing fibre optic Raman probes for applications in clinical spectroscopy". *Chem Soc Rev* 45(7): 1919–1934 (2016).

26. C Soliman, J Faircloth, D Tu, et al. "Exploring the clinical utility of Raman spectroscopy for point-of-care cardiovascular disease biomarker detection". *Appl Spectrosc* 77(10): 1181–1193 (2023).

27. Gar-Wing Truong, LW Perner, D Michelle Bailey, et al. "Mid-infrared supermirrors with finesse exceeding 400,000". *Nat Comm* 14: 7846 (2023).

28. RA Shaw, S Kotowich, HH Mantsch, M Leroux. "Quantitation of protein, creatinine, and urea in urine by near-infrared spectroscopy". *Clin. Biochem* 29(1): 11–19 (1996).

29. D Perez-Guaita, J Ventura-Gayete, C Perez-Rambla, et al. "Protein determination in serum and whole blood by attenuated total reflectance infrared spectroscopy". *Anal Bioanal Chem* 404(3): 649–656 (2012).

30. LM Rodrigues, TD Magrini, CF Lima, et al. "Effect of smoking cessation in saliva compounds by FTIR spectroscopy". *Spectrochim Acta A* 174: 124–129 (2017).

31. D Yonar, L Ocek, BI Tiftikcioglu, et al. "Relapsing-remitting multiple sclerosis diagnosis from cerebrospinal fluids via Fourier transform infrared spectroscopy coupled with multivariate analysis". *Sci Rep* 8(1): 13 (2018).

32. BR Wood, KR Bambery, MWA Dixon, et al. "Diagnosing malaria infected cells at the single cell level using focal plane array Fourier transform infrared imaging spectroscopy". *Analyst* 139(19): 4769–4774 (2014).

33. D Perez-Guaita, D Andrew, P Heraud, et al. "High resolution FTIR imaging provides automated discrimination and detection of single malaria parasite infected erythrocytes on glass". *Faraday Discuss* 187: 341–352 (2016).

34. H Kimura-Suda, T Ito. "Bone quality characteristics obtained by Fourier transform infrared and Raman spectroscopic imaging". *J Oral Biosci* 59(3): 142–145 (2017).

35. M Verdonck, A Denayer, B Delvaux, et al. "Characterization of human breast cancer tissues by infrared imaging". *Analyst* 141(2): 606–619 (2016).

36. L Novotny, N van Hulst. "Antennas for light". *Nat Photonics* 5: 83–90 (2011).

37. M Osawa, K Ataka, K Yoshi, Y Nishikawa. "Surface-enhanced infrared spectroscopy: The origin of the absorption enhancement and band selection rule in the infrared spectra of molecules adsorbed on fine metal particles". *Appl Spectrosc* 47: 1497–1502 (1993).

38. R Adato, H Altug. "In-situ ultra-sensitive infrared absorption spectroscopy of biomolecule interactions in real time with plasmonic nanoantennas". *Nat Commun* 4: 2154 (2013).

39. C Hughes, G Clemens, B Bird, et al. "Introducing discrete frequency infrared technology for high-throughput biofluid screening". *Sci Rep* 6: 20173 (2016).

40. KL Andrew Chan, SG Kazarian. "Attenuated total reflection Fourier-transform infrared (ATR-FTIR) imaging of tissues and live cells". *Chem Soc Rev* 45: 1850–1864 (2016).

41. H Cheon, H-J Yang, J-H Son. "Toward clinical cancer imaging using terahertz spectroscopy". *IEEE J Sel Top Quant Electr* 23(4): 1–9 (2017).

42. GP Gallerano, S Bierdon. "Overview of THz radiation sources". In *Proceedings of the 2004 FEL Conference*, pp. 216–221 (2004).

43. F Sizov, A Rogalski. "THz detectors". *Progress in Quant Electr* 34(5): 278–347 (2010).

44. AWM Lee, Q Qin, S Kumar, et al. "Real-time terahertz imaging over a standoff distance (>25 meters)". *Appl Phys Lett* 89(14): 141125 (2006).

45. A Dobroiu, M Yamashita, YN Ohshima, et al. "Terahertz imaging system based on a backward-wave oscillator". *Appl Opt* 43(30): 5637–5646 (2004).

46. TW Crowe, WL Bishop, DW Porterfield, et al. "Opening the terahertz window with integrated diode circuits". *IEEE J Solid-State Circ* 40(10): 2104–2110 (2005).

47. DB But, OG Golenkov, NV Sakhno, et al. "Silicon field-effect transistors as radiation detectors for the sub-THz range". *Semiconductors* 46, 678–683 (2012).

48. R Tauk, F Teppe, S Boubanga, et al. "Plasma wave detection of terahertz radiation by silicon field effects transistors: Responsivity and noise equivalent power". *Appl Phys Lett* 89(25): 253511 (2006).

49. S Ruffenach, A Kadykov, VV Rumyantsev, et al. "HgCdTe-based heterostructures for terahertz photonics". *APL Mater* 5(3): 035503 (2017).

50. AWM Lee, BS Williams, S Kumar, et al. "Real-time imaging using a 4.3-THz quantum cascade laser and a 320x240 microbolometer focal-plane array". *IEEE Photonics Technol Lett* 18(13): 1415–1417 (2006).

51. J-H Son, SJ Oh, H Cheon. Potential clinical applications of terahertz radiation. *J Appl Phys* 125(19): 190901 (2019).

52. H Cheon, HJ Yang, JH Son. "Toward clinical cancer imaging using terahertz spectroscopy". *IEEE J Sel Top Quantum Electron* 23: 1–9 (2019).
53. H Cheon, HJ Yang, SH Lee, et al. "Terahertz molecular resonance of cancer DNA". *Sci Rep* 6: 37103 (2016).
54. VP Wallace, AJ Fitzgerald, S Shankar, et al. "Terahertz pulsed imaging of basal cell carcinoma ex vivo and in vivo". *Brit J Derm* 151(2): s424–s432 (2004).
55. YB Ji, CH Park, H Kim, et al. "Feasibility of terahertz reflectometry for discrimination of human early gastric cancers". *Biomed Opt Express* 6(4): 1398–1406 (2015).
56. AJ Fitzgerald, VP Wallace, M Jimenez-Linan. "Terahertz pulsed imaging of human breast tumors". *Radiology* 239(2): 533–540 (2006).
57. MR Grootendorst, AJ Fitzgerald, SG Brouwer de Koning, et al. "Use of a handheld terahertz pulsed imaging device to differentiate benign and malignant breast tissue". *Biomed Opt Express* 8(6): 2932–2945 (2017).
58. K Okada, Q Cassar, H Murakami, et al. "Label-free observation of micrometric inhomogeneity of human breast cancer cell density using terahertz near-field microscopy". *Photonics* 8(5): 151 (2021).
59. YB Ji, SJ Oh, SG Kang, et al. "Terahertz reflectometry imaging for low and high grade gliomas". *Sci Rep* 6: 36040 (2016).
60. PA George, W Hui, F Rana, et al. "Microfluidic devices for terahertz spectroscopy of biomolecules". *Opt Express* 16(3): 1577–1582 (2008).
61. H Chen, X Chen, S Ma, et al. "Quantify glucose level in freshly diabetic's blood by terahertz time-domain spectroscopy". *J Infrared Milli Terahz Waves* 39: 399–408 (2018).
62. AM Fosnight, BL Moran, IR Medvedev. "Chemical analysis of exhaled human breath using a terahertz spectroscopic approach". *Appl Phys Lett* 103(13): 133703 (2013).
63. YB Ji, ES Lee, SH Kim, et al. "A miniaturized fiber-coupled terahertz endoscope system". *Opt Express* 17(19): 17082–17087 (2009).
64. YB Ji, IS Moon, HS Bark, et al. "Terahertz otoscope and potential for diagnosing otitis media". *Biomed Opt Express* 7(4): 1201–1209 (2016).
65. RM Woodward, VP Wallace, RJ Pye, et al. "Terahertz pulse imaging of ex vivo basal cell carcinoma". *J Invest Dermatol* 120(1): 72–78 (2003).
66. H Hoshina, A Hayashi, N Miyoshi, et al. "Terahertz pulsed imaging of frozen biological tissues". *Appl Phys Lett* 94(12): 123901 (2009).
67. SJ Oh, S-H Kim, K Jeong, et al. "Measurement depth enhancement in terahertz imaging of biological tissues". *Opt Express* 21: 21299–21305 (2013).

CHAPTER 10

Other Approaches in Physiological Optical Imaging

IN THIS CHAPTER, WE will discuss several other approaches that are used or can be used in the foreseeable future in clinical practice.

10.1 SPATIAL FREQUENCY DOMAIN IMAGING (SFDI)

Cuccia et al. [1] developed rapid, noncontact imaging for quantitative, wide-field characterization of turbid media's optical absorption and scattering properties based on spatial frequency domain imaging (SFDI). This group experimentally demonstrated that by projecting sinusoidal light patterns onto tissue, one could determine the tissue's optical properties by measuring the relative decay of spatial patterns at different frequencies. An algorithm was proposed for reconstructing three-dimensional images directly from measurements made by illuminating tissue with sinusoidal patterns [2] and experimentally proven as a quantitative optical tomography of subsurface heterogeneities (e.g., blood vessels) using spatially modulated structured light [3] and collecting images separated in terms of absorption and reduced scattering coefficients[4].

Once the spatial distribution of tissue's optical parameters (absorption coefficient and reduced scattering coefficient) have been established, they can be used to analyze tissue physiology and pathophysiology.

DOI: 10.1201/9781003482505-12

SFDI has been demonstrated preclinically to track wound healing in a diabetes model [5]. A clinical study has applied spatial frequency domain imaging (SFDI) to measure lower extremity perfusion and predict both the healing and formation of diabetic foot ulcers [6, 7]. McClatchy et al. [8] used SFDI to predict stromal, epithelial, and adipose fractions in freshly resected, unstained human breast specimens. They found that stromal, epithelial, and adipose volume fractions predicted from light scattering parameters strongly correlated with those calculated from digitized histology slides (r = 0.90, 0.77, and 0.91, respectively, p-value $< 1 \times 10^{-6}$).

10.2 POLARIZATION OPTICS

In addition to polarization filtration discussed in Chapter 3, polarization optics can retrieve additional information about the optical properties of tissues. In particular, several techniques are based on measuring the degree of polarization (or depolarization ratio), which is the quantity that describes the ratio of the intensity of polarized light to the total intensity of light, $P_L = (I_\parallel - I_\perp)/(I_\parallel + I_\perp)$, where $I_\parallel$ is the intensity of light polarized in parallel and $I_\perp$ perpendicular to the polarization plane. For example, polarization optics can characterize the optical anisotropy of fibrous structures, such as collagen and actin-myosin in soft tissues, and is commonly used in dermatology [9]. Tissue polarimetry was reviewed by Ghosh and Vitkin [10].

10.3 TISSUE MICROMOTION DETECTION

Several approaches try to exploit tissue motions caused by physiological processes. One particular example of such processes can be pulse propagation across major blood vessels.

While arterial pulse propagation is a well-known phenomenon, blood pulsations can be observed in certain veins. In particular, blood pulsations in the jugular vein have significant diagnostic values as they probe the right heart function [11].

Optical methods can detect blood pulsations, with pulse oximetry as the most well-known example. However, as blood pulsations represent distensions of major vessels, at certain conditions, these pulsations can be observed on the surface of the tissues as well.

To date, a few noncontact approaches have been proposed to measure the deformation of the jugular vein (JV) and the carotid artery (CA) as an indication of the blood pressure in the corresponding vessel. These methods assume that the deformation waveform shares features similar to those of the pressure waveform. Moco et al. [12] and HajiRassouliha et al. [13] have used noncontact techniques to measure JV or CA deformation waveforms. Using a color camera, Moco et al. [12] proposed a method to measure skin motion cardiac-related frequency components under nonuniform light, assuming that CA wall displacements dominate the motion. An alternative approach was proposed by HajiRassouliha et al. [13], who used a monochrome camera to measure skin deformations using subpixel image registration directly. Initially, this group developed a method for CA measurements [13]. Later, the technique has been extended to JV measurements [14].

The method of Moco et al. [12] required nonuniform illumination of the neck to increase variations in the frame-to-frame contrast of the moving regions.

A more traditional approach for JV waveform detection has been proposed by Amelard et al. [15]. This method uses photoplethysmography (PPG) imaging by a color camera. PPG imaging used by Amelard et al. [15] involves measuring the intensity of visible or infrared light reflected from the patient's neck to estimate blood vessel deformation waveforms caused by the JV. This technique typically requires a finger PPG device to identify the pulsatile waveforms related to the jugular vein (in addition to the neck PPG imaging device).

However, considering the depth of the internal jugular vein (1 or more cm) and the small sampling depth of remote PPG, the method is unlikely to sample the IJV itself. Several plausible approaches exist to interpret the signal's origin in this case. The first approach would be to think that mechanical forces caused by distensions of the jugular vein would modulate skin microcirculation, which rPPG can probe. The second line of reasoning can be that the rPPG signal, collected by the camera, contains a strong specular reflection component, which contains information about skin deformations caused by jugular vein distensions.

More recently, Saiko et al. [16] proposed to enhance skin micromotion detection using temporary skin marking (using a non-permanent marker pen) around target areas (e.g., internal jugular vein). Although it necessitates contact with the skin, the method simplifies instrumentation and data collection significantly.

Although, as discussed in section 3.5, specular reflection is a typically undesirable phenomenon that complicates physiological data extraction, there are attempts to use it for skin micromotion detection. In particular,

Specular Reflection Vascular Imaging has been proposed to collect carotid artery and jugular vein waveforms [17]. It explicitly exploits the specular reflection. For specular reflection to be maximal in a given angle of reflection, the surface must be as smooth as possible, per Figure 3.2b. As most skin surfaces in normal conditions can be considered rough, the smooth surface component of the specular reflection can be increased through the application of a "smoothening" agent, such as oil, that typically functions to populate the skin concavities (such as in Figure 3.2a) and generate a smooth surface (such as in Figure 3.2b). In line with this reasoning, the authors proposed treating the skin with oils to increase specular reflection [18].

Another approach to detecting small changes in the tissue surface shape would be holographic interferometry. This method is based on acquiring the holographic image of the object in the reference state. Then, the holographic interferogram can be obtained by reflecting the light from the deformed object on the holographic image. The analysis of such an interferogram can reveal important information about changes in the object's surface shape. In particular, Hariharan and Oreb [19] demonstrated that the method can reveal changes associated with changes in optical path length on the nanometer scale. The method has been intensively researched in otolaryngology. For example, Khaleghi et al. [20] developed spatial-bandwidth multiplexed holography for simultaneous full-field 3-D vibrometry of the human eardrum. They were able to measure magnitudes and phases of 3-D sound-induced motions of a human cadaveric tympanic membrane (the eardrum) at several excitation frequencies and observed modal and traveling wave motions on its surface.

REFERENCES

1. DJ Cuccia, FP Bevilacqua, AJ Durkin, et al. "Quantitation and mapping of tissue optical properties using modulated imaging". *J Biomed Opt* 14(2): 024012 (2009).
2. V Lukic, VA Markel, JC Schotland. "Optical tomography with structured illumination". *Opt Lett* 34(7): 983–985 (2009).
3. S Konecky, A Mazhar, D Cuccia, et al. "Quantitative optical tomography of sub-surface heterogeneities using spatially modulated structured light". *Opt Express* 17: 14780–14790 (2009).
4. S Gioux, A Mazhar, DJ Cuccia. "Spatial frequency domain imaging in 2019: Principles, applications, and perspectives". *J Biomed Opt* 24(7): 071613 (2019).
5. M Saidian, JRT Lakey, A Ponticorvo, et al. "Characterisation of impaired wound healing in a preclinical model of induced diabetes using wide-field

imaging and conventional immunohistochemistry assays". *Int Wound J* 16(1): 144–152 (2018).

6. GA Murphy, RP Singh-Moon, A Mazhar, et al. "Quantifying dermal microcirculatory changes of neuropathic and neuroischemic diabetic foot ulcers using spatial frequency domain imaging: A shade of things to come?" *BMJ Open Diab Res Care* 8(2): e001815 (2020).

7. https://clinicaltrials.gov/ct2/show/NCT03341559 as was visited on May 12, 2021.

8. DM McClatchy, EJ Rizzo, WA Wells, et al. "Light scattering measured with spatial frequency domain imaging can predict stromal versus epithelial proportions in surgically resected breast tissue". *J Biomed Opt* 24: 1–11 (2018).

9. RR Anderson. "Polarized light examination and photography of the skin". *Arch. Dermatol* 127: 1000–1005 (1991).

10. N Ghosh, IA Vitkin. "Tissue polarimetry: Concepts, challenges, applications, and outlook". *J Biomed Optics* 16(11): 110801 (2011).

11. MM Applefeld. "Chapter 19: The jugular venous pressure and pulse contour". In *Clinical Methods: The History, Physical, and Laboratory Examinations*, 3rd ed., HK Walker, WD Hall, JW Hurst, Eds. Boston: Butterworths (1990).

12. AV Moco, LZ Mondragon, W Wang, et al. "Camera-based assessment of arterial stiffness and wave reflection parameters from neck micro-motion". *Physiol Meas* 38: 1576–1598 (2017).

13. A HajiRassouliha, EJLP Tang, MP Nash, et al. "Quantifying carotid pulse waveforms using subpixel image registration". In *Computational Biomechanics for Medicine*, P Nielsen, A Wittek, K Miller, et al., Eds. Cham: Springer (2017).

14. EJ Lam Po Tang, A HajiRassouliha, MP Nash, et al. "Non-contact quantification of jugular venous pulse waveforms from skin displacements". *Sci Rep* 8(1): 17236 (2018).

15. R Amelard, R Hughson, D Greaves, et al. "Non-contact hemodynamic imaging reveals the jugular venous pulse waveform". *Sci Rep* 7: 1–10 (2017).

16. G Saiko, T Burton, Y Kakihana, et al. "Observation of blood motion in the internal jugular vein by contact and contactless photoplethysmography during physiological testing: case studies". *Biomed Opt Exp* 15(4): 2578–2589 (2024).

17. G Saiko, T Burton, A Douplik. "Feasibility of specular reflection imaging for extraction of neck vessel pressure waveforms". *Front Bioeng Biotechnol* 10: 830231 (2022).

18. T Burton, G Saiko, A Douplik. "Towards development of specular reflection vascular imaging". *Sensors* 22: 2830 (2022).

19. P Hariharan, BF Oreb. "Stroboscopic holographic interferometry: Application of digital techniques". *Opt Commun* 59: 83–85 (1986).

20. M Khaleghi, J Guignard, C Furlong, JJ Rosowski. "Simultaneous full-field 3-D vibrometry of the human eardrum using spatial-bandwidth multiplexed holography". *J Biomed Opt* 20(11): 111202 (2015).

PART 3

Endoscopy and Microscopy

SO FAR, WE HAVE considered physiological optical modalities in general. Most of them can be applied to large-area imaging. In this part of the book, we will consider endoscopic and microscopic techniques. These techniques are among the most commonly used in medicine. We devoted only one chapter per modality as these techniques are generally well-known. Thus, we decided to condense this information to a minimum and provide information mostly on recent developments.

Endoscopic and microscopic techniques are umbrella terms, as they can be combined with multiple other modalities, like fluorescence imaging or hyperspectral/multispectral imaging. However, we will take different approaches while discussing endoscopic and microscopic techniques. While we will discuss enhancements to traditional white light (WL) endoscopy like autofluorescence or narrow band imaging, for microscopy, we will narrow our discussion to microscopic techniques other than traditional wide-field microscopy (namely, confocal, light sheet, and two-photon microscopies) and *in vivo* applications of traditional wide-field microscopy.

DOI: 10.1201/9781003482505-13

CHAPTER 11

Endoscopic Physiological Optical Imaging

Endoscopy is a medical procedure to visualize the human body's internal organs or natural cavities [1]. It is one of the most applied imaging methods in medicine.

The endoscope is a rigid or flexible tubular device that covers diameters from the submillimeter range up to 25 mm, allowing a direct view into the body. Endoscopes are usually named according to the field of their application, organ, or procedure [2]. Currently, they include laparoscopy, thoracoscopy, gastrointestinal endoscopy (including esophagoscopy, gastroscopy, duodenoscopy, colonoscopy, and small bowel endoscopy), bronchoscopy, arthroscopy, urological endoscopy, gynecological endoscopy, ear, nose, and throat endoscopy, and neuroendoscopy [3].

Despite widespread applications, endoscopy mainly relies on white light (WL) imaging, which is considered "anatomical imaging" in our terminology and omitted from the discussion. However, several commonly used enhancements will be considered here.

11.1 CHROMOENDOSCOPY

Chromoendoscopy is an image-enhanced endoscopic technique that provides detailed contrast enhancement of the surface of the gastrointestinal mucosa. It is achieved either through dye-based chromoendoscopy or

DOI: 10.1201/9781003482505-14

electronic chromoendoscopy, which includes optical technologies such as flexible spectral imaging color enhancement (FICE, Fujinon) and i-scan (Pentax) [4].

11.1.1 Dye-Based Chromoendoscopy

Dye-based image-enhanced endoscopy (d-IEE) is an endoscopic technique that uses topical dyes such as methylene blue or indigo carmine applied during endoscopy to highlight differences in the mucosa, as well as dysplastic and malignant changes that are not apparent in white light.

The stains used for dye-based chromoendoscopy are divided into two major categories: absorptive stains (methylene blue) and contrast stains (indigo carmine) [4]. Methylene blue is absorbed by epithelial cells of the small or large intestine, which stain blue, unlike dysplastic and cancerous lesions, which remain unstained. Indigo carmine highlights mucosal topography by coating mucosal structures, pits, erosions, and depressions.

Dye-based chromoendoscopy is used to enhance mucosal irregularities. For example, it is often used to survey the esophagus for Barrett's esophagus, evaluate polyps in the colon, and examine dysplasia in inflammatory bowel disease (IBD) [4].

Dye-based chromoendoscopy is not routinely performed during general endoscopy as it requires specialized settings.

11.1.2 Electronic Chromoendoscopy

Flexible spectral imaging color enhancement (FICE) is a virtual chromoendoscopic technique developed by Fujifilm in 2005. FICE reconstructs the image and displays what the mucosa would look like if illuminated using a particular wavelength [5]. It is based on a spectral estimation technique that obtains a narrow bandwidth from a white light image without optical filters. The FICE processor generates many virtual wavelengths with increments of 5 nm.

Negreanu et al. [5] reviewed applications of FICE in digestive endoscopy and found that FICE provides better detection of lesions of "minimal" esophagitis, dysplasia in Barrett's esophagus, and squamous cell esophageal cancer. The use of FICE resulted in an improvement in the visualization of early gastric cancer. The authors also noted that current evidence does not support FICE for screening purposes in colon cancer, but it improves the characterization of colonic lesions.

11.2 NARROW-BAND IMAGING

Narrow-band imaging (NBI) aims to obtain information on mucosal and submucosal vasculature. In narrow-band imaging, the mucosa is illuminated by light with the help of narrow-band filters. In a typical scenario, the blue (415 nm) and green (540 nm) light is accentuated, as it is absorbed by hemoglobin for optimal detection of mucosal, glandular, and vascular patterns as well as the presence of abnormal blood vessels that are associated with the development of dysplasia. In this case, superficial mucosal vessels appear brown, while the thicker blood vessels in the submucosa appear cyan.

Boscolo Nata et al. [6] reviewed the utility of NBI in oncologic surgery across all medical specialties. They found that the primary utility of NBI is to highlight mucosal and submucosal vessels. Thus, the technology is useful for early detection, definition of resection margins, and follow-up.

For example, NBI is excellent for demonstrating the irregular microvessels of esophageal and gastric cancer at high magnification [7]. In gastroenterology, the narrow-band blue-green light improves the visualization of mucosal patterns; therefore, the color contrast is enhanced between the neoplastic lesions and adjacent normal mucosa [4].

Linked color imaging (LCI) and blue laser imaging (BLI) are further enhancements of the narrow band approach [8].

BLI was developed by Fujifilm (Japan) in 2011. It uses illumination by two laser sources. The first laser source (410 nm) produces a clear image of the superficial microvasculature of the digestive mucosa. Another laser source (450 nm) excites fluorescence, leading to the visualization of a deep vascular mucosal image. Combining these illuminations results in a new bright image of the digestive mucosa, including detailed microstructure and microvasculature [9].

Linked color imaging (LCI) was also developed by Fujifilm (Japan) in 2014. It uses a blue laser (410 nm) and a white light source. Unlike traditional NBI and BLI, where images are too dark in distant views to allow detection of superficial lesions, LCI provides sufficient light intensity with bright imaging even at a distant view [10]. As such, LCI contributes to detecting superficial lesions throughout the gastrointestinal tract by enhancing the color contrast between the neoplasm and the surrounding mucosa. LCI can distinguish them by their specific color allocation based mainly on the distribution of capillaries.

11.3 AUTOFLUORESCENCE IMAGING

Endoscopic autofluorescence imaging (AFI) produces real-time pseudo-color images based on natural tissue fluorescence generated from endogenous fluorophores (collagen, nicotinamide, adenine dinucleotide, flavin, and porphyrins). AFI can visualize lesions, including malignancies, by differences in tissue fluorescence properties and can reveal early-stage cancers not detectable by conventional white light endoscopy [11].

Endoscopic autofluorescence imaging (AFI) or autofluorescence endoscopy (AFE) is known under several names. For example, it was termed light-induced fluorescence endoscopy (LIFE) by Xillix (Canada), who brought this system to the US market in 1996.

The excitation wavelength of AFI is typically in the blue range. For example, in the LIFE system (Xillix, Canada), the He-Cd laser with a 442 nm excitation wavelength is used.

The clinical utility of autofluorescence imaging in cancer detection is based on findings that autofluorescence in neoplastic tissues is abnormal. In particular, it is reduced in green color (see Figure 11.1 for details) due to mechanisms mentioned in section 7.2.3. As a result, in a typical AFI image, normal tissue is pseudocolored as green and blood vessels as dark green. In contrast, hypertrophic fundic mucosa of the stomach and

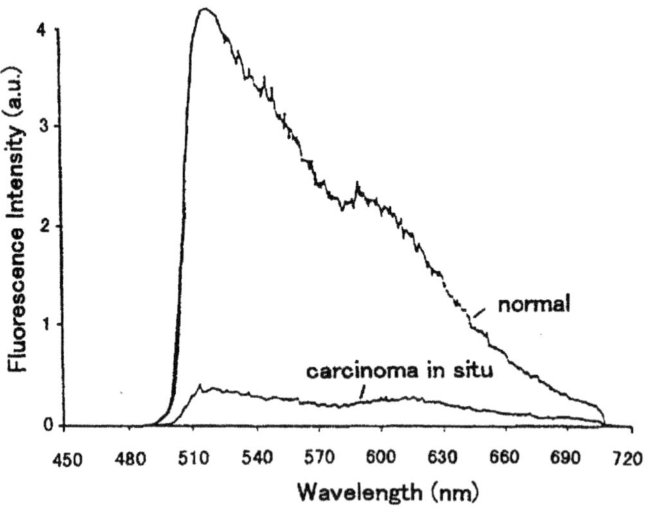

FIGURE 11.1 Autofluorescence spectrum of normal and abnormal tissue in the bronchus. Reproduced from [14] under CC BY 3.0 license.

dysplastic/neoplastic areas appear magenta [12]. Consequently, during AFI, a suspected neoplasia (AFI-positive lesion) is defined as a) any area that is different in color from the surrounding mucosa and b) has a defined circumferential margin [13].

AFI can be used with external agents. For example, Namihisa et al. [15] used the LIFE system in combination with noradrenaline spraying on patients with gastric cancers. They found that the normal mucosa became paler than the tumor after spraying noradrenaline, thereby clarifying the tumor boundary in the LIFE-GI image.

Bi et al. [16] reviewed the utility of AFI in diagnosing early gastrointestinal tumors. They found that gastrointestinal mucosal inflammation and proliferative lesions can also thicken the mucosal layer and show hues similar to tumor lesions under AFI, which affects the differential diagnosis of inflammatory lesions from tumor lesions. They concluded that compared with other special staining methods (such as NBI, chromoendoscopy, etc.), AFI fails to show a clear advantage in the accuracy of the diagnosis of gastrointestinal tumors [16].

In addition to traditional (single photon) autofluorescence endoscopy, some work has been done in two-photon microendoscopy (TPME). For example, rigid needlelike endoscopic GRIN lenses were used to image SHG in muscle sarcomeres below the skin surface [17]. However, due to the much more complex and expensive instrumentation (femtosecond lasers), TPME is fairly far from being translated and broadly used in endoscopy. Some details on two-photon microscopy and its clinical utility can be found in section 12.2.1.2.

11.4 PRACTICAL CONSIDERATIONS

Traditional WL endoscopic imaging has low sensitivity in detecting early-stage cancerous transformations. The proposed image enhancement techniques aim to address this shortcoming. However, they have their own limitations. For example, FICE remains somewhat poor at visualizing mucosal microvasculature on a tumor surface [9], which is a strong feature of blue laser imaging. Similarly, Ohkawa et al. [18] found that LIFE provided a high (96.4%) sensitivity but low (49.1%) specificity in stand-alone diagnostics of gastric lesions. Thus, the multimodal approach can significantly improve the sensitivity and specificity of endoscopy. For example, in the case of bronchoscopy, Yokomise et al. [19] reported that compared to conventional white-light bronchoscopy alone, the detection ratio of metaplasia and cancer was improved from 65% to 90%, and specificity was improved from 71% to 77.4% by using autofluorescence bronchoscopy combined with conventional bronchoscopy.

Fortunately, most of the current endoscopes provide some of these enhancements combined with traditional WL imaging. For example, oftentimes, it is recommended to use trimodal imaging, which includes autofluorescence imaging, magnifying endoscopy, and narrow-band imaging.

There are several trends in endoscopy. Firstly, there is a gradual shift from proximal cameras to chip-on-the-tip cameras. It introduces a lot of new environmental and technical requirements for cameras. Secondly, the new imaging modalities (which we discussed in previous chapters) are being added into consideration. For example, ICG-based NIR fluorescence endoscopy has been proposed for bile duct inspection [20]. Another obvious extension of imaging modalities is multispectral imaging [21, 22].

It should be noted that the clinical adoption of endoscopic technologies other than WL endoscopy is limited, at least in certain geographies. For example, in the United States, dye-based and electronic chromoendoscopy are used primarily in tertiary academic centers by expert endoscopists in high-risk patients for dysplasia detection and characterization and diminutive polyp management [4].

11.5 CONCLUSIONS

Endoscopy is a medical procedure to visualize the human body's internal organs or natural cavities. Despite widespread applications, endoscopy is still mainly limited to traditional white light imaging. It is caused, at least partially, by several factors. Firstly, with multiple proprietary technologies being developed, the knowledge is confined in silos of adaptors of these technologies.

Secondly, oftentimes, advanced endoscopic techniques demonstrate a small sensitivity and specificity. However, as they characterized tissue from a different angle, the multimodal approach can significantly improve the sensitivity and specificity of endoscopy.

REFERENCES

1. C Nezhat. *Nezhat's History of Endoscopy: A Historical Analysis of Endoscopy's Ascension Since Antiquity.* Miami, FL, USA: CNezhatMD (2011).
2. S Amornyotin. *Endoscopy.* London, UK: IntechOpen (2013).

3. A Boese, C Wex, R Croner, et al. "Endoscopic imaging technology today". *Diagnostics (Basel)* 12(5): 1262 (2022).

4. AM Buchner. "The role of chromoendoscopy in evaluating colorectal dysplasia". *Gastroenterol Hepatol (N Y)* 13(6): 336–347 (2017).

5. L Negreanu, CM Preda, D Ionescu, D Ferechide. "Progress in digestive endoscopy: Flexible spectral imaging colour enhancement (FICE)-technical review". *J Med Life* 8(4): 416–422 (2015).

6. F Boscolo Nata, G Tirelli, V Capriotti, et al. "NBI utility in oncologic surgery: An organ by organ review". *Surg Oncol* 36: 65–75 (2021).

7. M Kaise, M Kato, M Urashima et al. "Magnifying endoscopy combined with narrow-band imaging for differential diagnosis of superficial depressed gastric lesions". *Endoscopy* 41: 310–315 (2009).

8. J Weigt, P Malfertheiner, A Canbay, et al. "Blue light imaging and linked color imaging for the characterization of mucosal changes in chronic gastritis: A clinicians view and brief technical report". *Dig Dis* 38: 9–14 (2020).

9. H Osawa and H Yamamoto. "FICE and BLI". *Dig Endos* 26: 105–115 (2014).

10. S Shinozaki, H Osawa, Y Hayashi, et al. "Linked color imaging for the detection of early gastrointestinal neoplasms". *Therap Adv Gastroenterol* 12: 1756284819885246 (2019).

11. J Haringsma, GN Tytgat, H Yano, et al. "Autofluorescence endoscopy: Feasibility of detection of GI neoplasms unapparent to white light endoscopy with an evolving technology". *Gastrointest Endosc* 53: 642–650 (2001).

12. M Filip, S Iordache, A Săftoiu, T Ciurea. "Autofluorescence imaging and magnification endoscopy". *World J Gastroenterol* 17(1): 9–14 (2011).

13. M Kato, M Kaise, J Yonezawa, et al. "Autofluorescence endoscopy versus conventional white light endoscopy for the detection of superficial gastric neoplasia: a prospective comparative study". *Endoscopy* 39: 937–941 (2007).

14. S Takehana, M Kaneko, H Mizuno. "Endoscopic diagnostic system using autofluorescence". *Diagn Ther Endosc* 5(2): 59–63 (1999).

15. A Namihisa, H Miwa, H Watanabe, et al. "A new technique: Light-induced fluorescence endoscopy in combination with pharmacoendoscopy". *Gastrointest Endosc* 53(3): 343–348 (2001).

16. Y Bi, M Min, Y Cui, et al. "Research progress of autofluorescence imaging technology in the diagnosis of early gastrointestinal tumors". *Cancer Control* 28: 10732748211044337 (2021).

17. ME Llewellyn, RP Barretto, SL Delp, MJ Schnitzer. "Minimally invasive high-speed imaging of sarcomere contractile dynamics in mice and humans". *Nature* 454: 784–788 (2008).

18. A Ohkawa, H Miwa, A Namihisa, et al. "Diagnostic performance of light-induced fluorescence endoscopy for gastric neoplasms. *Endoscopy* 36(6): 515–521 (2004).

19. H Yokomise, K Yanagihara, T Fukuse, et al. "Clinical experience with lung-imaging fluorescence endoscope (LIFE) in patients with lung cancer". *J Bronchology* 4(3): 205–208 (1997).

20. WJ Tzeng, YH Lin, TY Hou, et al. "Near-infrared cholangiography can increase the chance of success in laparoscopic approaches to common bile duct stones, even with previous abdominal surgery". *BMC Surg* 23: 203 (2023).
21. L Qiu, R Chuttani, D Pleskow, et al. "Multispectral light scattering endoscopic imaging of esophageal precancer". *Light Sci Appl* 7: 17174 (2018).
22. S Ortega, H Fabelo, DK Iakovidis, et al. "Use of hyperspectral/multispectral imaging in gastroenterology. Shedding some–different–light into the dark". *J Clin Med* 8(1): 36 (2019).

Microscopic Physiological Optical Imaging

MULTIPLE MICROSCOPY-BASED TECHNIQUES ARE used in medicine. We will start with several *in vivo* applications of traditional wide-field microscopy. The wide-field microscopy is a two-dimensional (2D) technique. However, several microscopic techniques have been developed to allow three-dimensional (3D) imaging. They allow optical sectioning or slicing of the tissue. These techniques (confocal, light sheet, and two-photon microscopies) will be the primary focus of this chapter.

12.1 TRADITIONAL WIDE FIELD MICROSCOPY

Traditional wide field microscopy is the backbone of multiple clinical methods, particularly for *ex vivo* applications, including histology.

As the utility of wide-field microscopy is well understood, we will consider several *in vivo* applications.

12.1.1 Dermoscopy

Dermoscopy is the simplest optical modality, which provides 10x magnification of the area of interest. It is a noninvasive, inexpensive, handheld *in vivo* imaging technique with a diagnostic odds ratio > 15 times higher than that of naked eye examinations [1]. As such, it is a standard tool in

DOI: 10.1201/9781003482505-15

dermatologist's toolkits. Most modern devices use cross-polarized light to reduce the level of superficially reflected light from the skin surface and maximize visualization of deeper structures [2].

12.1.2 Videocapillaroscopy

Nailfold videocapillaroscopy (NVC) is a noninvasive diagnostic test used to study microvascular abnormalities. The technology uses the horizontal arrangement of nailfold capillaries, which allows their *in vivo* visualization. NVC allows the evaluation of the parameters of individual capillaries (length, shape, and diameter of each capillary loop, the number of capillaries) and the dynamic parameters (blood flow velocity) [3]. NVC is a simple, noninvasive, and reproducible test for adults and children [4]. Unlike dermoscopy, which uses 10x magnification, the typical magnification in capillaroscopy is around 200x.

The utility of videocapillaroscopy is well established for rheumatological disorders, such as connective tissue diseases (CTD).

NVC's most common diagnostic use is in the differentiation between primary and secondary Raynaud's phenomenon (RP), an intense vasospasm of the small arteries.

In addition, in systemic sclerosis (SSc), NVC allows detection of pathologic microvascular alterations and their activity phases, clustered in specific scleroderma patterns (early, active, and late) [5], which was evidenced by the inclusion of NVC scleroderma patterns into the new 2013 classification criteria for SSc.

In other CTDs, such as systemic lupus erythematosus, inflammatory myositis, undifferentiated connective tissue diseases, and mixed connective tissue disease, although nonspecific, the NVC patterns can provide valid support for the diagnosis [6].

The NVC use for non-rheumatological applications has been reviewed by Mansueto et al. [7]. They concluded that NVC could be a valuable tool for the early detection of endothelial dysfunction, which plays an important role in the pathogenesis of microvascular complications in several diseases, such as diabetes mellitus, pulmonary disease, hematological diseases, and CTDs. NVC can help discover new etiopathogenetic pathways in glaucoma, sickle cell disease, Werner syndrome, and Rett syndrome.

12.1.3 Dark Field Microscopy

Similar to orthogonal polarization imaging, several methods (typically contact modalities) exploit the idea of discarding surface reflectance. Dark field microscopy is one such modality. The Sidestream Dark Field (SDF) imaging device discards the surface-scattered photons by optically isolating the light guide from the illuminating outer ring. The SDF device comprises a central light guide in contact with the skin, surrounded by concentrically placed light-emitting diodes emitting green light. Goedhart et al. [8] developed and validated the SDF technology against OPS on nail-fold capillaries.

The technology subsequently evolved into the incident dark field (IDF) technique. The IDF is based on the incident dark-field illumination method initially developed by Sherman et al. [9]. Gilbert-Kawai et al. [10] compared IDF with SDF in sublingual microcirculatory imaging and found it superior.

12.2 3D MICROSCOPY

Three-dimensional microscopic techniques allow optical slicing or sectioning of the tissue. How they achieve it can be split roughly into three general approaches: focal spot microscopy (confocal microscopy, two-photon microscopy, and photothermal imaging), light sheet microscopy, and interferometric techniques (like OCT).

12.2.1 Focal Spot Microscopy

A group of high-resolution optical techniques that allows specimens to be monitored in multiple planes of depth by adjusting a specific focal depth can be referred to as focal spot microscopy. They are typically based on laser scanning technology.

12.2.1.1 Confocal Microscopy

The principles of confocal microscopy (CM) or laser scanning confocal microscopy (LSCM) were discussed earlier (see section 2.1.3.3). A number of fluorescence-based techniques are often combined with confocal microscopy, including fluorescence resonance energy transfer (FRET), fluorescence recovery after photobleaching (FRAP), fluorescence life-time imaging (FLIM), spectral imaging, optogenetics, and multiphoton imaging [11].

Reflectance confocal microscopy (RCM) is typically used for *in vivo* applications. In particular, RCM is a US FDA-approved optical imaging technology that offers noninvasive visualization of skin lesions *in vivo* at nearly histologic resolution. RCM allows for the simultaneous recording and evaluation of microcirculation, histomorphology, and inflammatory cell trafficking. For example, it has been used successfully to study burn wounds [12].

However, reflectance geometry has certain limitations. In particular, although the maximum imaging depth is about 300 μm, the imaging resolution decreases substantially below a depth of 100 to 150 μm, restricting accurate diagnostic interpretation to the epidermis and superficial dermis [13]. This depth limitation makes RCM less suitable for lesions that are nodular or have marked epidermal thickening, ulceration, or hyperkeratosis [14].

Confocal microscopy is used in clinical practice for diagnostics, treatment/management of skin cancers (delineating lateral tumor margins), and nonsurgical treatment monitoring [13].

In wound care, confocal microscopy has been used to image cutaneous wounds during healing noninvasively. In addition, it has been used to distinguish between normal and abnormal skin morphology [15]. However, currently, it is limited to superficial epidermal wounds [16].

12.2.1.2 Two-Photon Microscopy

Two-photon microscopy (TPM) is a microscopic technique widely used in cell biology, which can image biological structures with subcellular resolution. It has several significant advantages over other imaging techniques. Firstly, it uses NIR photons for excitation, allowing deeper sampling than other techniques. Secondly, as these NIR photons are less absorbed by the tissue and because high energy density is confined to the geometrical focus of the laser spot for a short period of time, it minimizes tissue photodamage. Finally, unlike traditional wide-field microscopy (e.g., epifluorescence), it allows 3D imaging of the tissue.

The TPM optical schema (laser-scanning type microscope) is quite similar to confocal microscopy. However, unlike confocal microscopy, a pinhole is not required, as all emitted photons convey a useful signal. Another big difference from confocal imaging is the light source. To obtain the high light intensities required to achieve two-photon (2P) fluorescence excitation, near-IR lasers typically produce very brief pulses (usually ~150 femtoseconds or less) at a high repetition rate (~100 MHz), which creates sufficient peak (instantaneous) energy to produce 2P excitation, while keeping average energy low enough to minimize specimen damage [17]. Mode-locked lasers are well suited to the requirements for efficient TPM. In particular, tunable femtosecond Ti: Al_2O_3 (titanium: sapphire) lasers are commonly used in most laboratories because the spectral range (690–1,050 nm) is sufficient to excite a wide variety of fluorophores [18].

TPM refers to two types of techniques: two-photon excitation fluorescence (TPEF) and second harmonic generation (SHG).

Two-photon excitation Fluorescence (TREF) is a fluorescence technique where excitation is achieved by the quasi-simultaneous absorption of two photons: the first one excites the electron to a virtual intermediate state, and the second one completes its excitation to reach the final excited state.

Second harmonic generation (SHG) does not include absorption of photons. Instead, SHG involves the scattering of photons and no energy loss, whereby two photons interacting simultaneously with a non-centrosymmetric target combine to produce a new photon with exactly twice the energy (thus, half the wavelength) of the interacting photons [19]. Unlike TREF, second harmonic generation is a coherent process. As such, it is sensitive to geometrical factors like scatterers' spacing, order vs. disorder, and orientation (see, for example, [20]), which can provide insights into molecular structure not readily available with TPEF alone.

Both TPEF and SHG are often referred to as "nonlinear" optical processes, as the intensity of their emission signals is proportional to the square of the irradiating intensity. While TREF and SHG use the same optical setup, they are characterized by different emission spectra (with and without energy loss, respectively). Thus, the only practical distinction between these techniques is different emission filters. As such, both techniques bear the same advantages. In particular, as a generated photon (either fluorescent in TREF or scattered in SHG) originates from the geometrical focus of the laser excitation spot, both TREF and SHG provide inherent optical sectioning.

As TREF microscopy can utilize either exogenous or endogenous (intrinsic) fluorescent molecules, TREF has broader applications in cell biology. Thus, the dominant fraction of works that use TPM is TREF based.

In addition to cell biology, the utility of TPM has been demonstrated in neuroscience and cancer pathology investigations.

The primary utility in neuroscience is neural structures imaging and recording of neural signals. In particular, the recording of neural signals can be achieved by either calcium imaging [21] or voltage-sensitive proteins [22]. Applications in neuroscience are primarily based on exogenous fluorescence. It can be achieved by filling cells with synthetic dyes, transgenic and gene targeting, and transfection methods [18].

In addition to preclinical cell and animal laboratory models, TPM has demonstrated its utility for investigating cancer pathology in *ex vivo* and *in vivo* human tissues [23]. Cancer research and clinical applications of TPM can be based on either exogenous or endogenous fluorescence. However, the primary advantage of TPM is the use of native fluorophores. TPEM of NADH and FAD[1], as well as both TPEM and SHG microscopy of collagen, are among the most investigated intrinsic fluorophores in cancer research [23].

NADH and $FADH_2$ (reduced FAD, i.e., flavin adenine dinucleotide hydroquinone) fluorescence (so-called "redox" ratio, see section 7.2.2) provides information about cells' metabolism and oxidative reduction capacity, which is particularly relevant to tumor growth and can be used to discriminate between cancerous vs. normal tissue.

Collagen is a highly ordered non-centrosymmetric protein that generates a strong SHG signal and a significantly weaker TPEF signal [23]. As collagen is a key component of the extracellular matrix (ECM), its utility of collagen in cancer diagnostics is based on theories that the organization/reorganization of collagen may play roles in tumor genesis [24], progression [25], and/or metastasis. For example, Nadiarnykh et al. [26] found that malignant ovarian tissue was characterized by denser and more ordered collagen and higher SHG intensity and scattering than normal tissue.

For *ex vivo* applications, TPM has been suggested for optical biopsy. For example, in *ex vivo* intact human gastrointestinal mucosa, Rogart et al. [27] found that using TPM with intrinsic fluorophores could discern a level of structural detail similar to that found in H&E slides and that TPM was superior to confocal microscopy. As TPM does not require sample preparation, it can be used in conjunction with current histological methods.

In addition to *ex vivo* applications, TPM can be used *in vivo*. In particular, TPM was applied on the skin and in microendoscopic applications. For example, Dimitrow et al. [28] used TPM of intrinsic TPEF to examine the pigmented skin cancer melanoma in 83 patients, both *ex vivo* and *in vivo*. They achieved an *in vivo* diagnostic sensitivity of 75% and an *in vivo* specificity of 80%.

12.2.1.3 Photothermal (PT) Imaging

Photothermal (PT) imaging is quite similar to photoacoustic imaging, which was discussed in Chapter 5. However, it has a substantial advantage as an all-optical technology, thus allowing subcellular resolution. This method has reached the sensitivity of single nanoparticles and single molecules. As such, it has multiple applications in biology and material science. The current status of PT and its applications have been reviewed by Adhikari et al. [29].

The photothermal signal arises from slight changes in the index of refraction in a sample caused by sample heating due to absorption of the (heating) beam. Then, the refractive index changes are measured with a second (probing) beam, usually of a different wavelength. Typically, two different strategies are used for the photothermal imaging of single nanoparticles: (i) photothermal interference contrast imaging [30] and (ii) photothermal heterodyne imaging [31].

Because of their fast nonradiative rates, photothermal imaging has shown great promise in detecting weakly or nonfluorescent molecular probes more resistant to photobleaching at higher excitation powers. Such molecules could be alternative molecular probes for biological applications, which could be small in size and highly photostable.

However, photothermal microscopy is currently limited to the transmission geometry, as its backward detection mode (i.e., reflection mode) is not well understood [29].

12.2.2 Light-Sheet Microscopy

Light-sheet microscopy (LSM) relies on an orthogonal separation of the illumination and detection path, enabling selective illumination of a whole imaging plane and simultaneous wide-field detection. LSM is typically used in fluorescence schema.

The primary requirement for LSM imaging is the need for a relatively transparent sample. As such, optical clearing techniques are typically used to increase the sample transparency [32].

Decoupling the light-sheet illumination from detection enables high-speed and large field-of-view imaging with minimal photobleaching and phototoxicity [33]. As such, LSM is uniquely positioned for histological applications. In particular, the optical sectioning capability of LSM can provide 3D volumetric information without physically cutting the specimen [34], reducing the biopsy time from hours to minutes.

Several groups adapted the technology to large samples. In particular, Glaser et al. [35] developed open-top light sheet microscopy (OT-LSM) and demonstrated it on large fresh human biopsy samples extracted from the breast $(2.0 \times 2.0 \times 0.4\,\text{cm})$, prostate $(3.1 \times 3.5 \times 0.4\,\text{cm})$, and kidney tissues. They demonstrated the utility of OT-LSM for various applications: wide-area surface microscopy to triage surgical specimens (with ~200 μm surface irregularities), rapid intraoperative assessment of tumor-margin surfaces $(12.5\,\mathrm{sec}/\mathrm{cm}^2)$, and volumetric assessment of optically cleared core–needle biopsies (1 mm in diameter, 2 cm in length). They also found that the OT-LSM images are comparable to standard histology images.

Similarly, Migliori et al. [36] developed light-sheet theta microscopy (LSTM) and demonstrated it on human brain sections of ~10.5×14.1×3 mm in size.

Abadie et al. [37] performed human skin biopsies using LSM. Five-millimeter-thick samples were optically cleared using benzyl-alcohol and benzyl-benzoate (Murray's clear method).

12.2.3 Interferometric Techniques (OCT)

Optical coherence tomography (OCT) is a well-established medical microscopic technique.

OCT is founded on the principle of low-coherence interferometry, which involves directing a light beam with low coherence (high bandwidth) onto the tissue being examined and then combining the scattered, back-reflected light with a second beam (known as the reference beam), which is obtained by splitting the original light beam. To retain coherence, OCT uses ballistic and near-ballistic photons. The concept of the OCT setup is depicted in Figure 12.1. The typical OCT setup includes a standard Michelson interferometer and a low time-coherence light source (e.g., a superluminescent diode or SLD).

There are many different implementations of OCT technology [38]. Two main OCT schemas are time-domain OCT (TD-OCT) and Fourier-domain OCT (FD-OCT). Fourier-domain OCT imaging can also be done in two ways: by spectral-domain OCT (SD-OCT) and swept-source OCT (SS-OCT).

Laterally adjacent depth scans (similar to the more familiar A-scans of ultrasound imaging technology) are used to obtain a sample's two-dimensional map of reflection sites. Note that there are two scan procedures in OCT. The reference mirror performs the OCT depth scan. The lateral OCT scan is performed by moving the sample or scanning the probe beam, illuminating the specimen.

Direct acquisition of 2D OCT images is accomplished through another method, full-field OCT (FF-OCT).

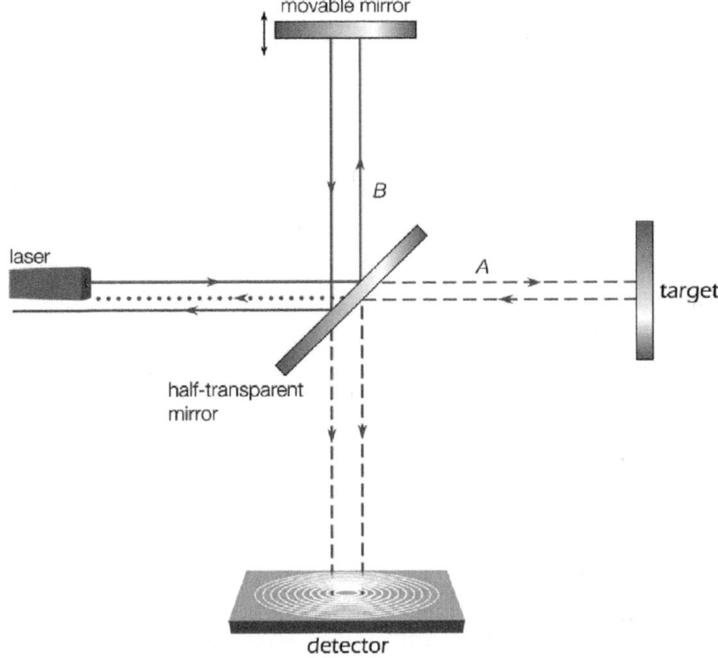

FIGURE 12.1   The concept of the OCT setup.

12.2.3.1 Clinical Utility

12.2.3.1.1 Ophthalmology   OCT was initially proposed as a technique for ophthalmology in 1991 [39]. Since then, it established clinical utility there. For example, optical coherence tomography angiography (OCT-A) is used for imaging the microvasculature of the retina and the choroid in humans.

OCT-A technology uses light reflectance on the surface of moving red blood cells to depict vessels accurately through different segmented eye areas. In addition to being noninvasive, OCT-A has several other advantages over other angiographic techniques: a) it has the capability of quantitative analysis of the retinal vessels, and b) it provides 3D imaging information of the macula and visualizes peripapillary capillaries that supply the retinal nerve fiber layer [40].

For example, OCT-A has been reported to be useful in the diagnosis and understanding of glaucoma [41] and many retinal conditions, including diabetic retinopathy [42], age-related macular degeneration [43], and vascular occlusions [44].

In recent years, several quantitative metrics measured by OCT-A have been developed [45], including vessel area density, vessel skeletal density, and vessel diameter index.

Several versions of OCT focus on the characterization of blood flow, particularly for ophthalmology. For example, OCT morphed with Doppler-based techniques into Doppler OCT (D-OCT), which aims to visualize and quantify blood flow. Traditional Doppler OCT can quantitatively measure blood velocity along the direction of the OCT beam; however, blood flow measurements require knowledge of the angle between the blood vessel and the light beam (the Doppler angle).

Several improvements to address this issue have been proposed. For example, blood vessels could be visualized using Doppler OCT with time domain OCT by extracting and comparing the A-scan phase changes related to the Doppler frequency shift in a method known as Optical Doppler Tomography [46]. Haindl et al. [47] proposed a three-beam Doppler optical coherence tomography (D-OCT) technique.

Now, the ability to provide flow information is a key feature of optical coherence tomography angiography (OCT-A). In particular, Braaf et al. [48] reviewed the utility of OCT-based velocimetry for retinal blood flow quantification.

12.2.3.1.2 Oncology OCT imaging has been used to image a broad spectrum of malignancies, including those arising in the breast, brain, bladder, gastrointestinal, respiratory, and reproductive tracts, as well as the skin and oral cavity, among others. OCT imaging has initially been applied for guiding biopsies, intraoperatively evaluating tumor margins and lymph nodes, and early detecting small lesions that would often not be visible on gross examination. Recently, OCT imaging has been explored for imaging tumor cells and their dynamics and monitoring tumor responses to treatments.

There are several primary scenarios where OCT has been contributing tremendously to the early detection of cancer [49]: (i) OCT can be used for cancer screening at higher resolutions than currently possible with the more established oncological imaging modalities; (ii) OCT can be used for guided biopsy, which is especially useful in situations where conventional biopsy is not effective, or where biopsies are associated with multiple complications; (iii) OCT can aid in the intraoperative imaging of cancer in the operation room by providing real-time feedback to the surgeons; (iv) OCT can be used for the monitoring of tumor responses to treatments.

All these scenarios arise from the fact that, unlike normal tissue with well-organized tissue structure, the structure of cancer tissue is often disorganized and typically characterized by variable cell sizes, abnormal shapes, and enlarged nuclei (see Section A.4.2), resulting in different optical scattering properties, which enables OCT to reveal differences between normal and cancerous tissues with high spatial resolution.

12.2.3.2 Practical Considerations

The spectral domain OCT (SD-OCT) uses light sources with a wavelength of around 800 nm, while a swept-source (SS-OCT) light source is close to 1.05 μm. Longer wavelengths have a deeper tissue penetration but a slightly lower axial resolution.

12.3 OTHER MICROSCOPIC TECHNIQUES

Several other contact imaging techniques have been proposed, including lensless microscopy [50], which uses multicore fibers, angular domain imaging [51], and numerical aperture-gated microscopy [52].

For example, angular domain imaging (ADI) uses high-aspect-ratio parallel microchannels (angular filter array or AFA) as an angular filter to capture quasibalistic photons [53]. The authors found that an angular filter with angular acceptance of about 0.4 deg yields the highest image contrast for transillumination images. The performance can be further improved by time-gating.

The use of a small numerical aperture (aperture-gated microscopy, [52]) has an advantage over ADI as it allows imaging in reflectance geometry.

12.4 CONCLUSIONS

Microscopic techniques in medicine and biomedical research can be split into two large groups: 2D and 3D microscopic techniques. Traditional wide field microscopy is a 2D microscopic technique. It is the backbone of multiple clinical methods, particularly for *ex vivo* applications, including histology. Three-dimensional microscopic techniques allow optical slicing or sectioning of the tissue. It is achieved using one out of three general approaches: focal spot microscopy (confocal microscopy, two-photon microscopy, and photothermal imaging), light-sheet microscopy, and interferometric techniques (like OCT).

Currently, these techniques are limited to *ex vivo* or areas with thin epidermis or mucosa.

NOTE

1 While NADH and FAD can be investigated by traditional one-photon microscopy, two-photon microscopy has an inherent advantage for *in vivo* applications as their absorption bands are in UV range, which is damaging to the tissue.

REFERENCES

1. ME Vestergaard, P Macaskill, PE Holt, SW Menzies. "Dermoscopy compared with naked eye examination for the diagnosis of primary melanoma: A meta-analysis of studies performed in a clinical setting". *Br J Dermatol* 159: 669–676 (2008).

2. C Benvenuto-Andrade, SW Dusza, AL Agero, et al. "Differences between polarized light dermoscopy and immersion contact dermoscopy for the evaluation of skin lesions". *Arch Dermatol* 143: 329–338 (2007).

3. W Grassi, R De Angelis. "Capillaroscopy: Questions and answers". *Clin Rheumatol* 26(12): 2009 (2007).

4. SS Ocampo-Garza, MA Villarreal-Alarcón, AV Villarreal-Treviño, J Ocampo-Candiani. "Capillaroscopy: A valuable diagnostic tool". *Actas Dermosifiliogr* 110(5): 347–352 (2019).

5. M Cutolo, S Paolino, V Smith. "Nailfold capillaroscopy in rheumatology: Ready for the daily use but with care in terminology". *Clin Rheumatol* 38(9): 2293–2297 (2019).

6. FP Cantatore, A Corrado, M Covelli, G Lapadula. "Morphologic study of the microcirculation in connective tissue diseases". *Ann Ital Med Int* 15(4): 273–281 (2000).

7. N Mansueto, C Rotondo, A Corrado, FP Cantatore. "Nailfold capillaroscopy: A comprehensive review on common findings and clinical usefulness in non-rheumatic disease". *J Med Invest* 68(1.2): 6–14 (2021).

8. PT Goedhart, M Khalilzada, R Bezemer, et al. "Sidestream Dark Field (SDF) imaging: A novel stroboscopic LED ring-based imaging modality for clinical assessment of the microcirculation". *Opt Exp* 15(23): 15101 (2007).

9. H Sherman, S Klausner, WA Cook. "Incident dark-field illumination: A new method for microcirculatory study". *Angiology* 22: 295–303 (1971).

10. E Gilbert-Kawai, J Coppel, V Bountziouka, et al. "A comparison of the quality of image acquisition between the incident dark field and sidestream dark field video-microscopes". *BMC Med Imaging* 16: 10 (2016).

11. SW Paddock, KW Eliceiri. "Laser scanning confocal microscopy: History, applications, and related optical sectioning techniques". In *Confocal Microscopy. Methods in Molecular Biology (Methods and Protocols)*, Vol. 1075, S Paddock, Ed. New York, NY: Humana Press (2014).

12. AA Altintas, M Guggenheim, MA Altintas, et al. "In vivo confocal laser scanning microscopy imaging of skin inflammation: Clinical applications and research directions". *J Burn Care & Res* 30(6): 1007–1012 (2009).

13. A Levine, O Markowitz. "Introduction to reflectance confocal microscopy and its use in clinical practice". *JAAD Case Rep* 4(10): 1014–1023 (2018).

14. C Longo, F Farnetani, S Ciardo. "Is confocal microscopy a valuable tool in diagnosing nodular lesions? A study of 140 cases". *Br J Dermatol* 169(1): 58–67 (2013).
15. M Rajadhyaksha, S Gonzalez, JM Zavislan, et al. "In vivo confocal scanning laser microscopy of human skin II: Advances in instrumentation and comparison with histology". *J Invest Derm* 113(3): 293–303 (1999).
16. S Lange-Asschenfeldt, A Bob, D Terhorst, et al. "Applicability of confocal laser scanning microscopy for evaluation and monitoring of cutaneous wound healing". *J Biomed Opt* 17(7): 076016 (2012).
17. BG Wang, K Konig, KJ Halbhuber. "Two-photon microscopy of deep intravital tissues and its merits in clinical research". *J Microsc* 238: 1–20 (2010).
18. R Mostany, A Miquelajauregui, M Shtrahman, C Portera-Cailliau. "Two-photon excitation microscopy and its applications in neuroscience". *Methods Mol Biol* 1251: 25–42 (2015).
19. PJ Campagnola, LM Loew. "Second-harmonic imaging microscopy for visualizing biomolecular arrays in cells, tissues and organisms". *Nat Biotechnol* 21: 1356–1360 (2003).
20. X Han, RM Burke, ML Zettel, et al. "Second harmonic properties of tumor collagen: Determining the structural relationship between reactive stroma and healthy stroma". *Opt Express* 16: 1846–1859 (2008).
21. C Grienberger, A Konnerth. "Imaging calcium in neurons". *Neuron* 73: 862–885 (2012).
22. A Perron, H Mutoh, W Akemann, et al. "Second and third generation voltage-sensitive fluorescent proteins for monitoring membrane potential". *Front Mol Neurosci* 2: 5 (2009).
23. SW Perry, RM Burke, EB Brown. "Two-photon and second harmonic microscopy in clinical and translational cancer research". *Ann Biomed Eng* 40(2): 277–291 (2012).
24. T Hompland, A Erikson, M Lindgren, et al. "Second-harmonic generation in collagen as a potential cancer diagnostic parameter". *J Biomed Opt* 13: 054050 (2008).
25. PP Provenzano, KW Eliceiri, JM Campbell, et al. "Collagen reorganization at the tumor-stromal interface facilitates local invasion". *BMC Med* 4: 38 (2006).
26. O Nadiarnykh, RB LaComb, MA Brewer, PJ Campagnola. "Alterations of the extracellular matrix in ovarian cancer studied by Second Harmonic Generation imaging microscopy". *BMC Cancer* 10: 94 (2010).
27. JN Rogart, J Nagata, CS Loeser, et al. "Multiphoton imaging can be used for microscopic examination of intact human gastrointestinal mucosa ex vivo". *Clin Gastroenterol Hepatol* 6: 95–101 (2008).
28. E Dimitrow, M Ziemer, MJ Koehler, et al. "Sensitivity and specificity of multiphoton laser tomography for in vivo and ex vivo diagnosis of malignant melanoma". *J Invest Dermatol* 129: 1752–1758 (2009).
29. S Adhikari, P Spaeth, A Kar, et al. "Photothermal microscopy: Imaging the optical absorption of single nanoparticles and single molecules". *ACS Nano* 14(12): 16414–16445 (2020).

30. D Boyer, P Tamarat, A Maali, et al. "Photothermal imaging of nanometer-sized metal particles among scatterers". *Science* 297: 1160–1163 (2002).

31. S Berciaud, D Lasne, GA Blab, et al. "Photothermal heterodyne imaging of individual metallic nanoparticles: Theory versus experiment". *Phys Rev B: Condens Matter Mater Phys* 73: 045424 (2006).

32. P Wan, J Zhu, J Xu, et al. "Evaluation of seven optical clearing methods in mouse brain". *Neurophotonics* 5: 035007 (2018).

33. PK Poola, MI Afzal, Y Yoo, et al. "Light sheet microscopy for histopathology applications". *Biomed Eng Lett* 9(3): 279–291 (2019).

34. OE Olarte, J Andilla, EJ Gualda, P Loza-Alvarez. "Light-sheet microscopy: a tutorial". *Adv Opt Photonics* 10: 111–179 (2018).

35. AK Glaser, NP Reder, Y Chen, et al. "Light-sheet microscopy for slide-free non-destructive pathology of large clinical specimens". *Nat Biomed Eng* 1: 0084 (2017).

36. M Migliori, MS Datta, C Dupre, et al. "Light sheet theta microscopy for rapid high-resolution imaging of large biological samples". *BMC Biol* 16: 57 (2018).

37. S Abadie, C Jardet, J Colombelli, et al. "3D imaging of cleared human skin biopsies using light-sheet microscopy: A new way to visualize in-depth skin structure". *Skin Res Technol* 24(2): 294–303 (2018).

38. DP Popescu, LP Choo-Smith, C Flueraru, et al. "Optical coherence tomography: Fundamental principles, instrumental designs and biomedical applications". *Biophys Rev* 3(3): 155 (2011).

39. D Huang, EA Swanson, CP Lin, et al. "Optical coherence tomography". *Science* 254(5035): 1178–1181 (1991).

40. Y Jia, JC Morrison, J Tokayer, et al. "Quantitative OCT angiography of optic nerve head blood flow". *Biomed Opt Express* 3(12): 3127–3137 (2012).

41. TE de Carlo, A Romano, NK Waheed, JS Duker. "A review of optical coherence tomography angiography (OCTA)". *Int J Retin Vitr* 1: 5 (2015).

42. RF Spaide. "Volume-rendered optical coherence tomography of diabetic retinopathy pilot study". *Am J Ophthalmol* 160(6): 1200–1210 (2015).

43. NK Waheed, EM Moult, JG Fujimoto, PJ Rosenfeld. "Optical coherence tomography angiography of dry age-related macular degeneration". *Dev in Ophthalmology* 6: 91–100 (2016).

44. T Wakabayashi, T Sato, C Hara-Ueno, et al. "Retinal microvasculature and visual acuity in eyes with branch retinal vein occlusion: Imaging analysis by optical coherence tomography angiography". *Investig Opthalmology Vis Sci* 58(4): 2087 (2017).

45. AY Kim, Z Chu, A Shahidzadeh, et al. "Quantifying microvascular density and morphology in diabetic retinopathy using spectral-domain optical coherence tomography angiography". *Invest Ophthalmol Vis Sci* 57: OCT362–OCT370 (2016).

46. Y Zhao, Z Chen, C Saxer, et al. "Phase-resolved optical coherence tomography and optical Doppler tomography for imaging blood flow in human skin

with fast scanning speed and high velocity sensitivity". *Opt Lett* 25: 114–116 (2000).

47. R Haindl, W Trasischker, A Wartak, et al. "Total retinal blood flow measurement by three beam Doppler optical coherence tomography". *Biomed Opt Express* 7(2): 287–301 (2016).

48. B Braaf, MGO Gräfe, N Uribe-Patarroyo, et al. "Chapter 7: OCT-based velocimetry for blood flow quantification". In *High Resolution Imaging in Microscopy and Ophthalmology: New Frontiers in Biomedical Optics [Internet]*, JF Bille, Ed. Cham (CH): Springer (2019). www.ncbi.nlm.nih.gov/books/NBK554059/ doi: 10.1007/978-3-030-16638-0_7

49. J Wang, Y Xu, SA Boppart. "Review of optical coherence tomography in oncology". *J Biomed Opt* 22(12): 1–23 (2017).

50. I Schelkanova, A Pandya, G Saiko, et al. "Spatially resolved, diffuse reflectance imaging for subsurface pattern visualization toward development of a lensless imaging platform: phantom experiments". *J Biomed Opt* 21(1): 015004 (2016).

51. F Vasefi, B Kaminska, GH Chapman, PKY Chan. "Angular domain imaging for tissue mapping". *2006 Bio Micro and Nanosystems Conference*, San Francisco, CA, USA, pp. 39–45 (2006).

52. A Pandya, I Schelkanova, A Douplik. "Spatio-angular filter (SAF) imaging device for deep interrogation of scattering media". *Biomed Opt Expr* 10(9): 4656–4663 (2019).

53. F Vasefi, M Najiminaini, E Ng, et al. "Angular domain transillumination imaging optimization with an ultrafast gated camera". *J Biomed Opt* 15(6): 061710 (2010).

CHAPTER 13

Future Directions in Physiological Optical Imaging

Throughout this book, we have debated the state of the art of diagnostic optical technologies in clinical practice. While demonstrating significant potential for diagnostics, most of them have not been translated into clinical practice yet.

There are several reasons for that. Firstly, technologies have just reached the level where biomedical applications have become a viable reality. Secondly, there is a significant mismatch between the supply and demand. In particular, not every market that seems obvious or lucrative is easily penetrable, and vice versa. It can be illustrated with cancer diagnostics in oncology. For example, scientists and engineers typically attack the most lucrative *in vivo* cancer diagnostics. However, breaking into the medical imaging market is difficult because current imaging modalities such as MR imaging and X-ray CT are widely utilized, and surgeons and radiologists depend heavily on them. The level of evidence needs to be overwhelming to convince them to switch from established imaging technologies to novel ones. As a more realistic strategy, alternative markets can be explored. For example, in the mentioned case of cancer diagnostics, as a simpler (and faster) step to clinical translation, researchers and engineers can consider *ex vivo* imaging for histology as the more viable commercial product.

DOI: 10.1201/9781003482505-16

In particular, surgeons currently judge the accuracy of tumor resection using H&E stained images, but these require at least a couple of days before the results can be observed. While some techniques, like freezing, can accelerate this process and bring quick decision-making in the operation room, they are not universally available. Thus, as different physiological optical imaging techniques (like vibrational and THz imaging) do not require sample preparation, they may become a method of choice for tumor diagnostics and margin determination with fast adoption.

In addition to technology advancements and the aforementioned supply/demand mismatch, several trends in medicine will likely drive the future development of optical medical modalities, particularly diagnostic ones. The remaining chapter briefly discusses them.

13.1 ARTIFICIAL INTELLIGENCE

Artificial intelligence (AI) is a field of computer science that focuses on creating machine systems that can perform tasks that typically require human intelligence. With the rapid penetration of AI into all aspects of human life, including the medical field, it is expected to be instrumental in the smart analysis of data derived from different sensors and technologies of relevance for screening, diagnosis, and care of various medical conditions.

In recent years, the medical field has experienced a significant influx of AI-based studies. Their exponential growth is driven primarily by the availability of open-access datasets.

Several branches of AI are in use or will be in use soon in the medical field.

13.1.1 Machine Learning

Machine learning (ML) refers to a specific subset of AI that focuses on developing algorithms and models that enable computers to learn and make predictions or decisions based on data without being explicitly programmed for every possible scenario. ML algorithms are designed to identify patterns and relationships within large data, allowing the system to generalize, make predictions, or act on new, unseen data.

For these reasons, broadly speaking, ML is used for five main tasks: automation, classification, regression, clustering, and prediction. Automation refers to the process of automating steps and tasks. It involves using computational systems to perform jobs that typically require manual intervention or human effort due to their repetitiveness, complexity, or attention to detail. By automating processes, ML tools and frameworks enable faster development, deployment, and scaling of products, help reduce manual effort, improve efficiency, enhance reproducibility, and enable humans to focus on higher-level tasks. The classification consists of assigning predefined categories or labels to input data based on its characteristics. ML algorithms can learn from labeled examples and then classify new, unlabelled data points into the appropriate categories.

Similarly, regression involves estimating a relationship between continuous numerical variables. ML algorithms can learn from input–output pairs to make predictions on new data. Finally, clustering is grouping similar data points based on their inherent patterns or similarities. In contrast to classification, which uses predefined labels, ML algorithms can automatically identify clusters or subgroups in data without predefined categories. These three techniques converge into prediction, which is by far the most common use of ML in clinical medicine.

ML algorithms can be categorized into supervised learning (where the algorithm learns from labeled data), unsupervised learning (where the algorithm discovers patterns in unlabeled data), and reinforcement learning (where the algorithm learns through trial-and-error interactions with an environment). The accuracy and reliability of predictions depend on various factors, including the quality and representativeness of the training data, the choice of the ML algorithm, the suitability of the features, and the model's generalization capability. Thus, continuous monitoring and updating of the model are essential to maintain accurate predictions as new data becomes available.

13.1.2 Deep Learning

Deep learning (DL) is a subfield of ML that focuses on training artificial neural networks (ANN) with multiple layers, hence the term "deep," to learn and make predictions or decisions based on large amounts of data. Deep learning is inspired by the structure and function of the human brain, particularly the optical cortex, where information processing occurs through interconnected neurons. Deep learning algorithms aim to automatically learn hierarchical representations of data by iteratively transforming the input through multiple layers of interconnected ANNs.

The key characteristics of deep learning include interconnected nodes or neurons organized into layers. These networks have an input layer, one or more hidden layers, and an output layer. The power of this is that in contrast to traditional machine learning approaches, where feature engineering is required to extract relevant features from the data, deep learning models can automatically learn representations of data from raw input. Deep learning models learn hierarchical representations, where lower layers capture simple features (e.g., edges or textures) and higher layers capture more complex and abstract features (e.g., object shapes or semantic concepts).

However, this complex behavior makes deep learning models act as "black boxes" where the steps or decisions required to devise a solution to the task get distributed between many interconnections. Furthermore, deep learning models are trained using an optimization algorithm called backpropagation. It involves computing the gradient of the model's error with respect to its parameters and adjusting the parameters to minimize the error. This process is repeated over multiple iterations or epochs until the model converges to an optimal solution. This backpropagation training further complicates the reproducibility of ANNs.

Nonetheless, deep learning has shown exceptional performance in various domains, including computer vision, natural language processing, speech recognition, recommendation systems, etc. In particular, it has achieved state-of-the-art results in tasks such as image classification, object detection, machine translation, sentiment analysis, and voice recognition.

Multiple ANN architectures have emerged in recent years. For example, recurrent neural networks (RNNs) use internal state (memory) to process inputs, making them suitable for unsegmented, connected handwriting recognition or speech recognition tasks. Convolutional neural networks (CNNs) are used for image analysis, including classification and clustering.

13.1.3 Generative AI

Generative AI is a recent addition to the AI field, which fascinated the public and is now the area of the arms race. Among several branches of generative AI, the more prominent are generative adversarial networks (GANs), which are used for image generation, and transformer models, which are used for text generation. In particular, GPT (generative pretrained transformer) is an example of the transformer model used for advanced text generation.

While GPT is a buzzword now, it is a sub-type of large language models or LLMs, which refer to any large-scale language model designed for natural language processing tasks. LLMs can use different architectures like transformers, recurrent neural networks (RNNs), and CNNs, offering scalability and versatility for various objectives.

The key feature of LLMs is that larger models predictively demonstrate better generalization across diverse datasets and scenarios. Thus, LLMs are characterized by extremely large sizes of these networks. For example, different versions of the LLaMA-2 (Large Language Model Meta AI) (Meta AI, USA) network launched by Meta AI in early 2023 contained 7, 13, 33, and 65 billion parameters. Even the smallest network, LLaMA 7B, was trained on one trillion tokens. Since then, model sizes have grown significantly.

LLM-based systems hold great promise in healthcare in the near future, particularly in patient management and chatbots. This promise is based on the approach to how these large language models are trained. It consists of two or three steps. In step one, the model is trained on a huge data set (e.g., the Internet). During this step, quantity is prioritized over quality. In step two, the model is fine-tuned using a much smaller number of high-quality resources (e.g., questions/answers provided by experts). So, at this stage, quality is prioritized over quantity. Thus, multiple medical domain LLM models may emerge based on the large pre-trained models. Particularly, it holds great promise with open LLM models like LLaMa (Meta AI, USA), as they allow fine-tuning large pre-trained models.

Thus, one can expect the emergence of medical domain AI chatbots. One particular example is MedPaLM (Google/DeepMind, USA).

The other likely trend will be the emergence of multimodal LLMs (M-LLM) for hospitals. These systems will be able to analyze not just text but image and video information and data from different clinical modalities.

13.1.4 Clinical Adoption of AI

While many AI-based approaches have been tried, the clinical adoption of AI-based technologies is still in its infancy. In addition to technology immaturity and small training sets, the primary obstacle right now is regulatory clearance. Like any other medical software, AI tools are medical devices regulated by market regulators like the US Food and Drug Administration (FDA). Until recently, regulators have taken a very cautionary approach to the clearance/approval of AI-based medical devices. For example, the software dhevelopment lifecycle for AI-based tools differs significantly from well-defined traditional medical software development. In particular, AI models aim to learn from new data. However, it contradicts traditional regulatory approaches where any changes to the medical device need to be properly managed. As such, regulators actively monitor the field and strive to develop a framework that will help streamline AI-based medical device development. For example, in October 2021, the US Food and Drug Administration (FDA), Health Canada, and the United Kingdom's Medicines and Healthcare Products Regulatory Agency (MHRA) jointly identified ten guiding principles that can inform the development of good machine learning practice (GMLP) [1].

Currently, the situation started changing. Even though the FDA database is not real time, as of winter 2023/2024, 692 AI-based medical devices have already been cleared by the FDA. Note that all these devices use traditional ML methods. For example, no devices with generative AI, like large language models (LLM), have been approved yet.

13.2 THERANOSTICS

New materials, including nanomaterials, have significantly opened up the field of theranostics, which refers to a field of medicine that combines diagnostics and therapeutics into a single integrated approach.

While being a buzzword for many years, theranostics has yet to demonstrate its clinical utility. As a relatively new concept, it is established just in several medical fields, particularly in nuclear medicine, where it couples diagnostic imaging and therapy using the same molecule or at least very similar molecules, which are either radiolabeled differently or given in different dosages [2].

However, it has significant potential in other medical fields. For example, one can think of nanoparticles, which can be used for both diagnostics/visualization and treatment of wounds or certain skin conditions. A prototype here can be porphyrins produced by certain bacteria, which can be used for both bacteria localization (at small light intensities) and as an antimicrobial photosensitizer (at higher light intensities).

Additionally, nanoparticles have a large surface area, which allows multiple functionalizations. In particular, multiple derivatization leads to combining different functionalizations, an indispensable feature for theranostics applications.

Thus, it is reasonable to expect that with nanomedicine developments, we will see many new applications, including theranostic ones.

Another interesting example stems from the THz field. DNA methylation, an epigenetic alteration independent of sequencing, initiates genetic mutations potentially leading to cancer [3]. Reversing methylation (or so-called demethylation) enhances cancer treatment by facilitating gene reprogramming, cell apoptosis, and tumor reduction [4]. Furthermore, combining demethylation with the antineoplastic agent decitabine significantly enhances chemotherapy efficacy [5]. Hence, there's potential to optimize cancer therapy by disrupting the methyl-DNA bond with THz radiation, an alternative to decitabine with notable clinical side effects [6]. Cheon et al. [7] demonstrated the possibility of demethylating various malignant DNAs using a high-power THz electromagnetic field. Thus, THz radiation at different intensities can be used for both methyl-DNA bond detection and demethylation.

13.3 MULTIMODAL DEVICES

The conventional approach to developing medical devices involves creating a device for a specific, often limited, purpose. Unfortunately, this has resulted in operating rooms, ICUs, and labs being filled with multiple devices that are frequently incompatible or unable to work together. Multimodal medical imaging involves using multiple imaging techniques or modalities to provide a more comprehensive and accurate assessment of a patient's condition. Combining several modalities into a single device is the natural approach to multimodality and offers significant benefits, particularly in image registration, synchronization, and increased specificity.

The benefits of a multimodal approach in medical imaging include improved visualization and understanding, enhanced image registration, synchronization of information, increased specificity and diagnostic accuracy, personalized treatment planning, and cost and space savings. By combining the strengths of different imaging modalities, healthcare professionals can provide a more comprehensive assessment and make better informed clinical decisions for optimal patient care.

13.4 INTEROPERABILITY: INTEGRATION WITH EHR SYSTEMS

Many medical devices are currently standalone, but the ability to integrate these devices into the electronic health record (EHR) system is becoming increasingly crucial. There are several benefits to integration:

1. Comprehensive Patient Information: Integration facilitates the seamless transfer of health care data and information between health care applications or devices and the EHR system. This ensures that wound-related information is integrated with the patient's overall medical record, providing a comprehensive view of their health status.

2. Care Coordination and Continuity: Integration with EHR systems supports better care coordination among healthcare providers involved in the patient's treatment. It allows different healthcare professionals to access and contribute to the patient's care data in a coordinated manner, promoting continuity of care and reducing the risk of duplicate or conflicting treatments.

3. Improved Clinical Decision-making: Integration with EHR systems gives healthcare providers a more holistic and accurate picture of the patient's health. Access to complete health care data and other clinical information helps make informed decisions about treatment plans, assessment, and interventions.

4. Efficiency and Time Savings: Integration reduces the need for manual data entry and duplication of efforts. Patient data captured through mobile devices or specialized applications can be automatically transferred to the EHR system, saving time and reducing the potential for errors in data entry.

5. Research, Advanced Algorithms, and Reporting: Integrated healthcare data in EHR systems can be used for research, training of advanced algorithms, quality improvement initiatives, and reporting purposes.

Integration with EHR systems ensures patient data is accessible, shared, and utilized effectively, leading to better care coordination, improved decision-making, and enhanced patient outcomes.

There are several levels of integration with EHR systems: no integration, generic integration, and full integration. No integration refers to cases where the captured information is either not stored or stored locally on the device. Generic interoperability protocols like HL7 or FHIR provide basic interoperability between various EHR systems (generic interoperability). Full integration with a particular EHR system offers significant functionality advantages. However, how is it possible to integrate a medical device with a multitude of existing EHR systems? The answer is that there is probably no need to do it. If the EHR market had been fragmented ten years ago, then by the time of writing, the market would have become much less fragmented, with clear winners emerging in so-called verticals (e.g., hospitals or home healthcare). For example, Epic (Epic Systems, Wisconsin, US) is the de facto EHR system for hospitals in the US. Thus, for any given target market (or vertical), certain EHR systems can be targeted first to get the most coverage.

13.5 WIDE USE OF SMARTPHONES

Smartphones have the potential to transform patient care by providing convenient and accessible methods for disease assessment, monitoring, and communication between healthcare providers and patients. Smartphones can significantly reduce healthcare costs using built-in high-resolution cameras and additional modalities like NIR or thermal imaging. Here's how smartphones can be used in wound care:

1. Anatomical Wound Imaging: Smartphones with high-resolution cameras can capture detailed wound images, which can be securely transmitted to healthcare providers for remote assessment and tracking of wound progression over time.

2. Wound Measurement Apps: Specialized smartphone apps can accurately measure wound dimensions using built-in cameras and measurement algorithms, aiding in documenting wound size and tracking healing progress without specialized equipment.

3. Physiological Wound Imaging: Smartphones paired with other devices can create sophisticated medical tools, enabling multiple imaging modalities such as HSI/MSI and fluorescence imaging.

4. Telemedicine and Virtual Consultations: Smartphones enable virtual consultations, allowing patients to share real-time images or videos of their wounds with healthcare providers for remote assessment and recommendations.

5. Wound Care Management Apps: Dedicated smartphone apps offer features like dressing change reminders, medication schedules, wound care instructions, and tracking wound healing progress, helping patients and caregivers stay organized.

6. Patient Education and Self-management: Smartphone apps provide educational resources on wound care, including videos, diagrams, and instructions, enabling patients to learn about proper wound care techniques and self-management.

7. Remote Monitoring and Alerts: Smartphones can connect to wearable sensors or smart dressings to monitor parameters at the wound site and transmit data to healthcare providers in real time for early detection of potential issues or complications.

While smartphones offer numerous benefits in healthcare, maintaining patient privacy and data security is essential. Adherence to relevant regulations and guidelines is crucial to protect patient information when utilizing smartphones in healthcare.

13.6 TELEMEDICINE AND REMOTE MONITORING: HEALTHCARE CLOSER TO PATIENT

Oftentimes, patients present to healthcare providers at later stages of the disease, when more serious interventions are required. If healthcare can be moved closer to the patient, the disease can be detected earlier, thus improving outcomes and the quality of the patient's life. Optical modalities can be integrated into telemedicine and remote monitoring, thus bringing diagnostics and disease management closer to the patient. For example, telemedicine technologies can remotely monitor wounds or skin conditions, provide consultation, and facilitate follow-up care, reducing the need for frequent in-person visits.

In particular, chronic wounds are traditionally managed through hospital or clinic appointments with multiple specialists such as wound care physicians, podiatrists, infectious disease specialists, vascular surgeons, and wound care nurses. Chronic wounds are typically seen weekly for several months before completely healing. Predictably, outcomes for chronic wounds are substantially worse for patients residing in remote areas with limited access to specialists. In rural areas, amputations are up to ten times higher than in urban areas.[8]

Thus, telehealth greatly appeals to patients in rural areas and those with transport barriers in urban areas.

The telemedicine (TM) solution helps save time by reducing the frequency of visits to wound clinics, which was reported from Australian and Canadian sources where patients living in distant rural areas had to take several days off to attend a consultation. Furthermore, patients living in rural areas also reported the benefit of avoiding stress related to driving in large, unfamiliar cities, looking for parking spaces, eating places, and toilet facilities.[9]

Patient-focused studies find patients did not want their relatives or friends to be obliged to help. Patients would rather have an equal and normal relationship with their family and friends. The TM solutions increased their experience of independence, contributing to decreased guilt of being a burden.[10] Also, expenses related to the workplace, where they had to miss a day's work, often with the result that they missed a day's pay.[11]

In many countries, different modalities of TM have already been implemented as a communication tool between caregivers across the healthcare sectors. These implementations demonstrated a greater feeling of responsibility and awareness for signs of infection by patients.[12]

As such, there is significant interest in telemedicine among patients, providers, and payers as a useful vehicle to supplement traditional face-to-face care. However, the cost-effectiveness of telehealth management is not fully understood. Published financial studies are fragmented and narrow in scope and only explore economics from a singular perspective, e.g., the provider, patient, hospital, etc. To date, no "total cost accounting" study has considered all stakeholders over a long period of time.

In summary, while it is true that not all health care can or should be delivered virtually, it is equally true that not all health care must occur in the clinic.

REFERENCES

1. www.fda.gov/medical-devices/software-medical-device-samd/good-machine-learning-practice-medical-device-development-guiding-principles (accessed Jun 25, 2023).
2. A Yordanova, E Eppard, S Kürpig, et al. "Theranostics in nuclear medicine practice". *Onco Targets Ther* 10: 4821–4828 (2017).
3. JS You, PA Jones. "Cancer genetics and epigenetics: Two sides of the same coin?" *Cancer Cell* 22(1): 9–20 (2012).
4. T Nakaoka, Y Saito, H Saito. "Aberrant DNA methylation as a biomarker and a therapeutic target of cholangiocarcinoma". *Int J Mol Sci* 18(6): 1111 (2017).
5. PM Howell, Z Liu, HT Khong. "Demethylating agents in the treatment of cancer". *Pharmaceuticals (Basel)* 3(7): 2022–2044 (2010).
6. HM Kantarjian, S O'Brien, J Cortes, et al. "Results of decitabine (5-aza-2′deoxycytidine) therapy in 130 patients with chronic myelogenous leukemia". *Cancer* 98(3): 522–528 (2003).
7. H Cheon, JH Paik, M Choi, et al. "Detection and manipulation of methylation in blood cancer DNA using terahertz radiation". *Sci Rep* 9(1): 6413 (2019).
8. A Drovandi, S Wong, L Seng, et al. "Remotely delivered monitoring and management of diabetes-related foot disease: An overview of systematic reviews". *J Diab Scie Techn* 17(1): 59–69 (2023).

9. SF Søndergaard, EG Vestergaard, AB Andersen, et al. "How patients with diabetic foot ulcers experience telemedicine solutions: A scoping review". *Int Wound J* 20: 1796–1810 (2023).

10. C Boodoo, JA Perry, PJ Hunter, et al. "Views of patients on using mHealth to monitor and prevent diabetic foot ulcers: Qualitative study". *JMIR Diab* 2(2): e22 (2017).

11. J Devine. "User satisfaction and experience with a telemedicine service for diabetic foot disease in an Australian rural community". In *Greater Southern Area Health Service (GSAHS) online Bega Valley Community Health*. Hujenni: Greater Southern Area Health Service (GSAHS) (2007).

12. B Ploderer, R Brown, LSD Seng, et al. "Promoting self-care of diabetic foot ulcers through a mobile phone app: User-centered design and evaluation". *JMIR Diab* 3(4): e10105 (2018).

Appendix A: Surface Tissues Morphology

The appendix's primary goal is to provide an overview of the structural properties of the surface tissues, namely skin and mucosa, which are important in medical practice, particularly for diagnostics. In addition, we provide a brief overview of pathomorphology in the case of wounds and carcinomas.

A.1 SKIN

The skin is the human body's largest organ, covering approximately 2 sq. m. It plays a vital role in protection and thermoregulation.

A.1.1 Skin Types

The skin can be subdivided into glabrous (non-hairy) and non-glabrous (hairy) skin. Glabrous (or hairless) skin is found on the palms of the hands, soles of the feet, and fingertips. As the primary point of contact with the external environment, glabrous skin differs from other skin types. In particular, it is much thicker and more durable. Additionally, glabrous skin is a major organ for sensing the external environment (somatosensation). As such, it is more sensitive to touch due to the high density of sensory receptors in this area.

Non-glabrous or hairy skin is found on the rest of the body. It is the prevalent skin type covering approximately 90% of the body surface. The presence of hair follicles characterizes non-glabrous skin.

The epidermis is less than 0.1 millimeters thick in hairy skin, and the dermis is 1–2 millimeters deep. Glabrous skin is thicker than hairy skin; the epidermis is about 1.5 millimeters thick, and the dermis is about 3 millimeters deep.

A.1.2 Skin Morphology

The skin is divided into three major layers: epidermis, dermis, and subcutaneous tissue. Figure A.1 depicts the composition of skin.

A.1.2.1 Epidermis

The epidermis is the outermost layer of the skin, which protects the body from external insults such as UV radiation and physical trauma. The epidermis comprises two sublayers, namely the nonliving and living epidermis. It consists of four types of cells, namely keratinocytes (which produce keratin), melanocytes (which produce melanin), Langerhans cells, and Merkel cells. Among these cells, keratinocytes are the most abundant. They are arranged in five distinct layers (stratums), namely the stratum corneum, stratum lucidum, stratum granulosum, stratum spinosum, and stratum basale, each with a specific function.

The outermost layer of the epidermis is referred to as the stratum corneum. This layer comprises flattened, dead skin cells that have migrated to the skin's surface. This layer acts as a barrier to water loss and helps to protect the body from external insults.

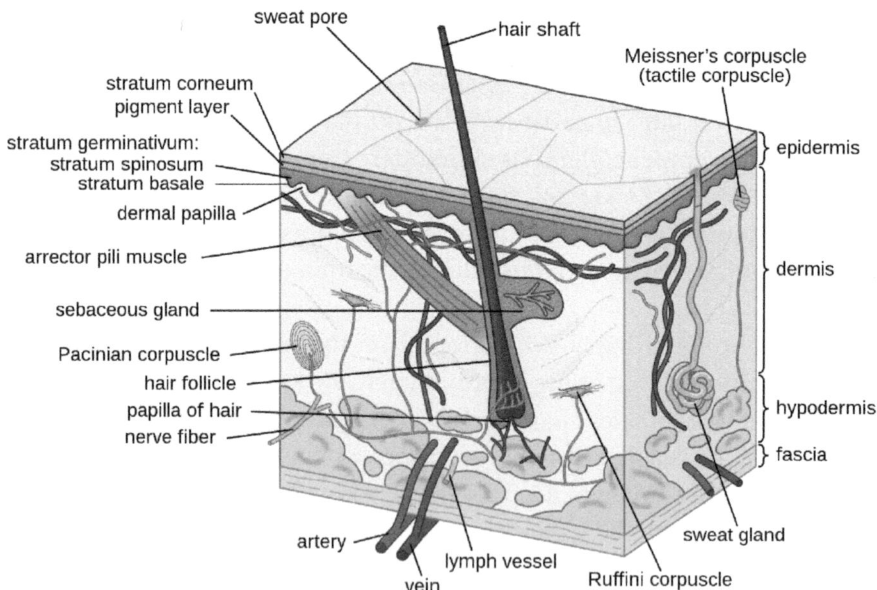

FIGURE A.1   Skin morphology. (Source: Tomáš Kebert & umimeto.org under CC BY 4.0 license).

Below the stratum corneum is the stratum lucidum, found only in glabrous skin. It is also composed of dead cells and packed with lipid-rich eleidin, which helps to keep water out. The stratum corneum and stratum lucidum form the nonliving epidermis. The nonliving epidermis is composed entirely of dead squamous cells that are highly keratinized and contain high amounts of lipids (~20%) and proteins (~60%) [1]. This layer is usually around 20 micrometers thick and has relatively low water content, comprising only about 20% of its composition.

The living epidermis (~100 μm thick) consists of living cells produced in continually dividing basal cells, which push older cells upwards.

Below the nonliving epidermis is the stratum granulosum, composed of dying and becoming flattened keratinocytes. As they mature, keratinocytes produce keratin, which helps to strengthen the skin and protect it from damage.

The stratum spinosum is the next layer of the epidermis, composed of actively dividing cells. This layer primarily consists of keratinocytes held together by desmosomes.

The basal cell layer, also known as the stratum basale, is the deepest layer of the epidermis. It is composed of stem cells responsible for continuously producing new skin cells. As these cells mature and move toward the skin's surface, they differentiate into the various cell types found in the other epidermis layers. A basal lamina separates the epidermis from the dermis.

In addition to the production of keratinocytes, the stratum basale contains melanin, which is responsible for most skin pigmentation and protects skin from the harmful effects of UV radiation. Melanin is produced in melanocytes and is found in membrane-bound intracellular organelles called melanosomes. There are two types of melanin: phaeomelanin, which appears reddish yellow, and eumelanin, which appears brownish black. The amount of melanin in the skin, which depends on the number of melanosomes present per unit volume of the epidermis, affects its ability to absorb light. In particular, the proportion of the epidermis occupied by melanosomes varies, ranging from 1.3% in individuals with light skin pigmentation to 43% in those with dark skin pigmentation [2].

A.1.2.2 Dermis

The dermis is the inner layer of the skin, located below the epidermis. It is a thick and resilient layer of tissue that provides structural support to the skin, sensory function, and immune protection.

The dermis comprises two main layers: the papillary and reticular layers. The papillary layer is the dermis's outer layer, composed of loose, connective tissue that contains small blood vessels, nerve endings, and immune cells. This layer provides nutrients and oxygen to the epidermis cells and detects sensations such as touch and temperature.

The reticular layer, which is the deeper layer of the dermis, is mainly made up of dense connective tissue containing collagen and elastin fibers. These fibers provide strength and elasticity to the dermis, enabling it to stretch and move without tearing. The reticular layer also contains larger blood vessels, sweat glands, and sebaceous glands, which are responsible for maintaining the body's temperature and moisture levels.

The dermis also contains various specialized cells, such as fibroblasts, which produce collagen and elastin fibers, and macrophages, which help defend the body against infection and disease. The dermis is a layer of the skin that contains a rich network of blood vessels, which gives it its characteristic pinkish color. The main absorbers in the visible spectral range in this layer are hemoglobin, carotene, and bilirubin. In the near-infrared spectral range, the absorption properties of the dermis are mainly defined by the amount of water present in it. The dermis also has a fibrous structure, mainly composed of collagen fibers that give it strength and elasticity. These fibers scatter light differently, depending on their size and arrangement. The papillary layer has smaller collagen fibers that scatter light in multiple directions, while the reticular layer has larger collagen fibers that scatter light mainly in a forward direction. The amount of blood and water also varies between the two layers, with the papillary layer having a higher blood volume fraction and the reticular layer having a higher water content. On average, the volume fraction of water in the dermis is estimated to be around 70%.[3] Blood content can be assessed from 0.2% [3] to 4%.[4]

A.1.2.3 Subcutaneous Tissue

The subcutaneous tissue, also known as the hypodermis or subcutis, is the layer of tissue located beneath the dermis and above the underlying muscle and bone. It is a layer of fat and connective tissue that serves several critical bodily functions, including insulation, energy storage, and protection of the underlying tissues.

One of the main functions of the subcutaneous tissue is to provide insulation and energy storage. The fat cells in this layer, called adipocytes, store excess energy in triglycerides, which can be mobilized when the body needs additional fuel. The subcutaneous tissue also helps regulate body temperature by insulating and preventing heat loss.

The subcutaneous tissue also acts as a cushion, protecting the underlying tissues from physical trauma. In particular, it helps to distribute the forces of impact evenly across the body, protecting the bones and internal organs from damage.

Subcutaneous tissue contains blood vessels, nerves, and lymphatic vessels in addition to fat cells. These structures provide nutrients and oxygen to the cells, help remove waste products, and maintain immune function.

Adipose tissue is composed of fat cells containing stored lipids in the form of droplets. There are multiple small droplets in each cell for lean or normal humans, while for obese humans, there is usually a single large drop. These spherical droplets are evenly distributed within adipocytes. The diameters of the adipocytes range from 15 µm to 250 µm, with a mean diameter of 50 µm to 120 µm. In the spaces between the cells, there are blood vessels, nerves, and reticular fibrils that provide metabolic activity.

A.2 MUCOSA

The mucosa, or the mucous membrane, is the soft tissue that covers the canals and organs within the digestive, respiratory, and reproductive systems of the body. It has three layers: the epithelium, lamina propria, and muscularis mucosae.

A.2.1 Epithelium

The surface layer of the mucosa is known as the epithelial layer. These cells produce a thick, gel-like mucus that protects the body from irritants. The arrangement of these cells varies to suit different body parts, with one or several layers of cells stacked in columns or laid like bricks. Additionally, these cells have a high turnover rate, constantly replacing themselves and clearing out invasive particles. Some of these cells also feature tiny hairs known as cilia, which aid in removing foreign substances.

A.2.2 Lamina Propria

The mucosal layer contains a connective tissue called the lamina propria, which attaches to the epithelium. It comprises structural protein molecules, nerves, and veins, supplying blood to the epithelium and holding the cells in place. The nerves in the lamina propria respond to muscle movements, altering the shape of the epithelium when necessary. Additionally, it houses immune cells that work to eliminate pathogens.

A.2.3 Muscularis Mucosae

The innermost layer of the mucosa is called the muscularis mucosae, which comprises smooth muscle. Its thickness varies in different parts of the digestive tract, and it is most active in the stomach. The muscularis mucosae keep the mucosa in motion, allowing it to stretch and contract along with the digestive organs. Additionally, it may help cleanse the skin by keeping the cilia on the surface in motion.

A.3 CUTANEOUS MICROCIRCULATION

Human whole blood comprises 55 vol% plasma (90% water, 10% proteins) and 45 vol% cells (99% red blood cells, 1% leukocytes, and thrombocytes). Red blood cells (RBC) or erythrocytes are the primary components of the blood. The RBC has a characteristic flat biconcave form with a diameter of 7 to 8 μm and a thickness of 2 μm. RBCs are smaller than other cells, including white blood cells (12 μm). This small size allows RBCs to fit easily through the microvasculature. Each erythrocyte contains approximately 260 million hemoglobin molecules, which occupy 95% of its volume [5].

The microvasculature is a network of small blood vessels responsible for providing oxygen and nutrients to the cells in various human body organs and removing waste products. The organization of the microvasculature in the skin is consistent across different parts of the body and age groups, with only slight variations in the density of capillary loops and ascending arterioles. The main components of the microvasculature in the skin are capillary loops, an upper plexus located in the papillary dermis, a lower plexus located at the dermal–subcutaneous interface, and arterioles and venules connecting them. The upper plexus is a horizontal network of capillaries that supplies nutrients to the dermal papillae. In contrast, the lower plexus gives rise to ascending arterioles that connect to the upper plexus. Most of the microvasculature in the skin is located in the papillary dermis, just 1–2 millimeters below the surface of the epidermis. The arterioles in the papillary dermis vary from 17 to 26 μm in diameter and represent terminal arterioles.

The capillaries in the skin, which are the smallest blood vessels, originate from a terminal arteriole in the upper layer of the dermis. They form a loop-like structure of an ascending branch, a horizontal section with a hairpin turn, and a descending branch that connects to a postcapillary venule in the lower layer of the dermis. The diameter of the capillaries is relatively small, with an outer diameter of 10–12 μm and an inner endothelial tube diameter of 4–6 μm [6].

The branching pattern of the arterioles that supply blood to the capillaries is not uniform and occurs at random intervals of 1.5–7 mm [7]. Each arteriole branches into four to five smaller branches that form the upper horizontal plexus of blood vessels. The blood vessel pattern in this network can be considered an umbrella shape, where the paired ascending arteriole and descending venule represent the handle. This "umbrella," comprising arterial and venous microvessels, is often called a "vascular unit." The vascular unit can be considered the smallest unit of microcirculation. For example, one would expect that the body regulates microcirculation on the vascular unit level.

The ultrastructure of the arterioles and venules in the reticular layer differs from the ultrastructure of comparable vessels in the superficial horizontal plexus. In particular, their diameters are wider, 50 μm versus 25 μm, and the walls are thicker, 10–16 μm versus 4–5 μm [8]. The arterioles and venules in the adipose tissue are identical in structure and size to those of the lower horizontal plexus.

The body has various mechanisms to control the blood flow in small blood vessels of microcirculation. These mechanisms include the autonomic nervous system, metabolic processes, and muscle contractions. However, the primary function of microcirculation regulation is to control body temperature (thermoregulation).

There are some variations in microcirculation between hairless and hairy skin. In hairless skin, there are connections between arterial and venous compartments called arteriovenous anastomoses (AVAs). In hairy skin, a ring of smooth muscle cells surrounds the branching point of the small blood vessels, forming a sphincter-like structure. The network of small blood vessels, called arterioles, plays a key role in determining the amount and distribution of blood flow to the capillaries.

A.4 PATHOMORPHOLOGY

A.4.1 Wounds

If the integrity of the skin has been breached (abrasions, lacerations, punctures, or avulsions), the wound-healing processes start. Normally, they occur in a sequence: homeostasis, inflammation, proliferation, and maturation. However, the normal wound-healing cycle can be disrupted.

Therefore, a chronic wound can be defined as physiologically impaired due to a disruption of the wound-healing process due to impaired angiogenesis, innervation, or cellular migration, among other reasons. The wound is considered chronic if it is not healed within three (or four, by some sources) weeks. The Wound Healing Society classifies chronic wounds into four major categories: pressure ulcers, diabetic foot ulcers, venous ulcers, and arterial insufficiency ulcers.

If the integrity of the skin or mucosa is broken, it becomes an acute wound. In this case, some underlying tissues like bones, ligaments, or nerves can be exposed. Some additional tissues (so-called wound tissues) can be present as the wound healing occurs. This section will provide some basic information on tissues that may be present in the presence of a wound.

Several factors can influence a wound's appearance, including the wound's size and depth, the presence of foreign bodies and other tissue types, such as slough or eschar (dead tissue). One of the key factors used to classify wounds is the stage of the healing process. There are four stages of wound healing: hemostasis (bleeding control), inflammation, proliferation (tissue growth), and maturation (tissue repair). The appearance of a wound can vary significantly depending on which stage of the healing process it is in. For example, a wound in the hemostasis stage may appear fresh and bleeding, while a wound in the maturation stage may appear as a fully healed scar.

Other factors that can affect the appearance of a wound include the presence of infection and the presence of other tissue types. An infected wound may appear red, swollen, and painful and may have a foul smell. A wound that contains slough or eschar may appear yellow or tan and may have a wet or dry appearance. To properly assess and manage a wound, it is important to carefully examine its morphology and consider the various factors influencing its appearance, which can help to identify any potential issues or complications and guide the selection of appropriate treatment options.

A.4.1.1 Wound Tissue Types

Several types of wound tissue can be present in a wound. These include:

A.4.1.1.1 Epithelial Tissue The epithelium is light pink and shiny, pearl-like. Epithelial cells migrate from the outer edges of the wound and move across the wound bed until the wound is closed.

A.4.1.1.2 Granulation Tissue   Granulation tissue is a pink, spongy tissue formed during healing. It comprises new blood vessels, collagen, and other cells. Granulation tissue helps to fill in the wound and provides a framework for the growth of new skin.

A.4.1.1.3 Necrotic (Nonviable) Tissue   Necrotic (nonviable) tissue is the tissue that has died due to a lack of blood flow. Necrotic tissue can be black, brown, or yellow in color and may have a foul smell. Removing necrotic tissue from a wound prevents infection and promotes healing.

Eschar is a thick, blackened layer of dead tissue that forms over a wound. It is often the result of a burn or a severe infection.

A.4.1.1.4 Slough   Slough is a yellow or greenish-yellow coagulum often found in wounds with a lot of drainages. It comprises dead white blood cells and is a sign of infection or tissue death. As such, it is not truly a tissue.

Nonviable tissue is often used collectively to describe different types of necrotic tissue. For example, necrotic tissue that undergoes hardening and turns black is known as hard black eschar. As autolysis progresses, the texture and color of the tissue change, resulting in the formation of a soft brown eschar, which eventually transforms into either an adherent grey/yellow slough or a loosely adherent slough. Sometimes, necrotic tissue includes eschar, but slough is considered a different tissue type.

Proper wound care involves regularly inspecting and cleaning the wound to remove any dead or infected tissue and promote the growth of new, healthy tissue, which can help to accelerate healing and reduce the risk of complications.

A.4.1.2 Other Tissues

Some other tissues can be exposed in the case of the wound. They include muscles, nerves, and various types of connective tissues: tendons, ligaments, cartilage, bones, and adipose tissue.

A.4.1.2.1 Muscles   Muscles are vital in supporting the skeletal structure and facilitating body movement. Healthy striated muscles are characterized by their bright red color due to the abundance of blood vessels present throughout the muscle tissue. Additionally, when touched or palpated, these muscles have a firm and elastic feel.

The smallest structural and functional subunits of skeletal muscle are myofibrils that lie parallel to the long axis of the muscle fiber. The diameter of myofibrils is about 1–2 µm, and the resting length of a sarcomere is about 2–3 µm long [9].

The basic units of the skeletal muscles are muscle fibers. The diameter of a mature skeletal muscle fiber ranges from 10 to 100 µm. Three layers of connective tissue protect skeletal muscle: the epimysium, the outermost layer, encircles the entire muscle; the perimysium surrounds groups of 10 to 100 muscle fibers, separating them into bundles called fascicles; and the endomysium separates individual muscle fibers from one another.

The optical properties of muscle depend on measurement orientation due to the anisotropic nature of light propagation in muscle. For example, the relative difference in the measured absorption coefficients between the 0° and the 90° probe orientations is as much as 50%.[10].

A.4.1.2.2 Nerves   Nerves are cordlike structures that are responsible for transmitting stimuli between the central nervous system and different parts of the body. They are composed of bundles of myelinated or unmyelinated nerve fibers and contain connective tissue. Nerves are part of the peripheral nervous system, which is different from the central nervous system (brain and spinal cord). They are grayish white in color and are located within the subcutaneous tissue layer, making them susceptible to unintended damage.

A.4.1.2.3 Adipose Tissue   Adipose tissue, commonly known as fat, is primarily located in the subcutaneous layer of the skin and typically appears as a globular, yellowish-white mass (except for newborns, when it may be brown in color). However, if the adipose tissue becomes damaged, its color may darken to a deeper shade of yellow.

A.4.1.2.4 Bones   Bone is a rigid and strong connective tissue that makes up the body's skeleton. It comprises cells called osteocytes, surrounded by a hard matrix of minerals. When exposed, bones appear white or pale yellow in color and have a hard texture when healthy. Bones are often overlooked during visual assessments, so it may be necessary to gently probe the wound base with the wooden end of a cotton-tipped applicator or palpate the area with a gloved finger to confirm the presence of hard bone.

A.4.1.2.5 Tendons and Ligaments   Dense connective tissue is found in areas of the body where there is a need for strength and stability, such as tendons and ligaments. It comprises closely packed collagen or elastic tissue fibers and is typically less flexible than loose connective tissue. Tendons and ligaments connect two other tissues together (i.e., bone to muscle or bone to bone, respectively). Their color can range from white to yellow, depending on the hydration level. Healthy tendons are shiny and white. However, when damaged or dehydrated, it darkens to a yellowish color as it begins to die. Therefore, maintaining moisture in tendons is crucial.

A.4.1.2.6 Callus   Corns and calluses are thickened areas of skin caused by repeated friction, pressure, or irritation. They form when dead skin cells accumulate due to the skin's response to protect itself from irritation.

The formation of calluses involves both biochemical and mechanical changes in the skin. First, mechanical stress activates keratinocytes in the epidermis, which divide and differentiate faster, forming a thicker layer of keratin. This protein makes up the stratum corneum. In addition, mechanical stress leads to the buildup of extracellular matrix molecules like collagen and glycosaminoglycans, contributing to the thickening of the stratum corneum.

Risk factors, such as foot deformities (e.g., bunions, hammertoe) and not wearing socks or protective gloves, can contribute to callus formation. Although calluses are generally harmless, if left untreated, they can result in skin ulceration or infection, particularly concerning for individuals with diabetes. They can also cause patients to shift their weight away from the affected, painful area, leading to excessive stress on the asymptomatic side.

Therefore, callus removal is an essential part of surgical debridement. However, healthcare professionals often have difficulty identifying calluses, which are often unsightly and not as clearly visible as corns. So, some areas of dead skin can be missed during the debridement process.

A.4.2 Carcinomas

There are more than 200 types of cancer. They can be classified in two ways: by the type of tissue in which the cancer originates (histological type) and by the primary site or the location in the body where the cancer first developed [11]. According to the International Classification of Diseases for Oncology, Third Edition (ICD-O-3) [12], there are six main groups:

- Carcinoma – this cancer arises out of epithelial cells that line the surface and organs of the body.

- Sarcoma – this cancer begins in the connective or supportive tissues such as bone, cartilage, fat, muscle, or blood vessels.

- Lymphoma – Lymphomas develop in the glands or nodes of the lymphatic system.

- Leukemia – "blood cancers" are cancers of lymphocytes and myelocyte precursors. It can be found in blood and bone marrow.

- Myeloma – a cancer that originates in the plasma cells of bone marrow.

- Mixed types – can be within one category or from different categories like adenosquamous carcinoma

Carcinomas are much more common than other types of cancers, accounting for 85% to 90% of all cancer cases [13]. In contrast, sarcomas represent slightly less than 1% of cancer types [13]. The remaining percentage includes other types of cancers, such as leukemias, lymphomas, and myelomas.

According to ICD-O-3, carcinomas are divided into two major subtypes [11]: adenocarcinoma, which develops in an organ or gland, and squamous cell carcinoma, which originates in the squamous epithelium. However, in practice, carcinomas are typically classified into four subtypes:

- "Adenocarcinoma – Adenocarcinomas start in glandular cells called adenomatous cells, which produce fluids to keep tissues moist.

- Basal cell carcinoma – Basal cells line the deepest layer of skin cells. Cancers that start in these cells are called basal cell carcinomas.

- Squamous cell carcinoma – Squamous cell carcinoma starts in squamous cells. These are the flat, surface-covering cells found in areas such as the skin or the lining of the throat or esophagus.

- Transitional cell carcinoma – Transitional cells are cells that can stretch as an organ expands. They make up tissues called transitional epithelium. An example is the lining of the bladder. Cancers that start in these cells are called transitional cell carcinoma."

Carcinomas, a type of cancer originating in epithelial tissues, can exhibit several typical changes or characteristics. These changes may include:

- Loss of Normal Tissue Organization: Carcinomas typically disrupt the normal organization of epithelial tissues. Normal cells are arranged in an organized, structured manner, but this arrangement is often lost in carcinomas. Cells may become disorganized and chaotic in their growth pattern.

- Hyperplasia: Carcinomas often involve an increase in the number of cells in the affected tissue, a condition known as hyperplasia, which leads to an enlargement of the affected area.

- Dysplasia: Dysplasia refers to abnormal changes in the size, shape, and appearance of cells within the tissue. In carcinomas, cells may show varying degrees of dysplasia, with some cells appearing markedly different from normal cells.

- Nuclear Changes: The nuclei of carcinoma cells frequently exhibit abnormalities. These changes can include enlarged nuclei, irregularly shaped nuclei, and increased chromatin (genetic material) within the nucleus.

- Increased Mitotic Activity: Carcinoma cells often divide more rapidly than normal cells, resulting in an increased number of mitotic figures (cells undergoing division) when observed under a microscope.

- Invasion and Infiltration: One hallmark of carcinomas is their ability to invade surrounding tissues. Carcinoma cells can break through the basement membrane (a barrier that separates epithelial tissue from underlying tissue) and infiltrate neighboring structures.

- Loss of Cell Adhesion: Normal epithelial cells are held together by cell adhesion molecules. In carcinomas, these molecules can be disrupted, leading to a loss of cell adhesion and the ability of cancer cells to detach from the primary tumor site.

- Formation of Tumor Mass: Over time, carcinoma cells can clump together to form a tumor mass. This mass may vary in size and shape depending on the type and stage of the carcinoma.

- Angiogenesis: Carcinomas often stimulate the formation of new blood vessels in a process called angiogenesis, which provides the growing tumor with a blood supply, allowing it to receive nutrients and oxygen.

- Metastasis: One of the most critical morphological changes associated with carcinomas is the ability of cancer cells to break away from the primary tumor, enter the bloodstream or lymphatic system, and spread to distant organs or tissues. This process is known as metastasis and is a hallmark of malignant carcinomas.

Some of these features can be detected/analyzed using optical means. Optical properties of various types of malignant tissues were analyzed in [14].

While most of the mentioned features can be used for diagnostics and staging, we will consider some common features used in optical diagnostic techniques. In particular, we will focus on the loss of normal tissue organization and angiogenesis, considered hallmark cancer development features.

A.4.2.1 Loss of Normal Tissue Organization

Carcinomas exhibit a distinctive feature known as the loss of normal tissue organization. This phenomenon is characterized by significant deviations from healthy tissues' orderly arrangement and structure. Several key factors contribute to this disruption:

- Cell Adhesion Dysfunction: In most cases, cancer cells become less adhesive than their healthy counterparts. This change often results from reduced expression of cell surface adhesion molecules, making it difficult for cancer cells to adhere tightly to neighboring cells [15].

- Polarity Disruption: Loss of apical–basal polarity is a hallmark of epithelial cancers, occurring early in tumor progression. Epithelial cells in carcinomas lose their characteristic polarity, leading to disorganized cellular architecture [16].

- Uncontrolled Growth: Carcinoma cells exhibit unregulated growth and proliferation, breaking the balance between cell division and cell death. This uncontrolled cell division contributes to the formation of tumor masses and further disrupts normal tissue structure [17].

- Invasive Behavior: Cancer cells acquire the ability to invade surrounding tissues, leading to the destruction of normal tissue architecture. This invasive behavior results from the loss of tissue organization and the capacity to breach basement membranes [15].

The loss of normal tissue organization is critical to carcinoma development and progression. This phenomenon is broadly used in optical diagnostic techniques that provide structural and morphological information, such as optical coherence tomography (OCT).

A.4.2.2 Angiogenesis

Angiogenesis, the process of new blood vessel formation, plays a pivotal role in the progression of carcinomas. In carcinomas, tumors require an adequate blood supply to fuel their growth and metastasis.

Tumor angiogenesis was reviewed by Lugano et al. [18]. Tumors can be vascularized either through co-option of the pre-existing vasculature or by inducing new blood vessel formation through various molecular and cellular mechanisms [18]. Angiogenesis is tightly regulated by a complex interplay of signaling pathways involving angiogenic factors, such as vascular endothelial growth factor (VEGF), fibroblast growth factor 2 (FGF2), and platelet-derived growth factors (PDGF).

As carcinomas develop, tumor cells release pro-angiogenic factors that stimulate nearby blood vessels to sprout and invade the tumor microenvironment. These new blood vessels supply oxygen and nutrients to the growing tumor mass, facilitating its expansion and survival. Moreover, angiogenesis promotes tumor cell dissemination by providing routes for metastatic spread to distant organs.

For example, hepatocellular carcinoma (HCC) is a hypervascular solid tumor characterized by an aberrant vascular network. Angiogenesis contributes significantly to its growth, invasion, and metastasis. As HCC expands, the oxygen and nutrient supply becomes insufficient beyond a certain lesion diameter (usually 1–2 mm) [19].

As angiogenesis marks more advanced stages of cancer progression, optical means can detect associated structural and functional (metabolic) changes.

REFERENCES

1. GF Odland. "Structure of the skin". In *Physiology, Biochemistry, and Molecular Biology of the Skin*, Vol. 1, LA Goldsmith, Ed. Oxford: Oxford University Press, p. 362 (1991).
2. SL Jacques. "Origins of tissue optical properties in the UVA, visible and NIR regions". In *Advances in Optical Imaging and Photon Migration*, Vol. 2, RR Alfano, JG Fujimoto, Eds. Washington, DC: OSA TOPS: Optical Society of America, p. 364371 (1996).

3. AN Bashkatov, EA Genina, VV Tuchin. "Optical properties of skin, subcutaneous, and muscle tissues: A review". *J Innov Opt Health Sci* 4(1): 9–38 (2011).

4. IV Meglinski, SJ Matcher. "Quantitative assessment of skin layers absorption and skin reflectance spectra simulation in visible and near-infrared spectral region". *Physiol Meas* 23: 741753 (2002).

5. T Yoshida, M Prudent, A D'alessandro. "Red blood cell storage lesion: Causes and potential clinical consequences". *Blood Transfus* 17(1): 27–52 (2019).

6. IM Braverman. "The cutaneous microcirculation". *J Invest Dermatol* 5(1): 3–9 (2000).

7. IM Braverman, J Schechner. "Contour mapping of the cutaneous microvasculature by computerized laser Doppler velocimetry". *J Invest Derrnatol* 97(6): 1013–1028 (1991).

8. IM Braverman, A Keh-Yen. "Ultrastructure of the human dermal microcirculation. III. The vessels in the mid- and lower dermis and subcutaneous fat". *J Invest Dermatol* 77: 297–304 (1981).

9. HJ Swatland. "Relationship between pork muscle fiber diameter and optical transmittance measured by scanning microphotometry". *Can J Anim Sci* 82: 321–325 (2002).

10. G Marquez, LV Wang, S Lin, et al. "Anisotropy in the absorption and scattering spectra of chicken breast tissue". *Appl Opt* 37: 798–804 (1998).

11. National Cancer Institute. https://training.seer.cancer.gov/disease/categories/classification.html

12. World Health Organization. www.who.int/standards/classifications/other-classifications/international-classification-of-diseases-for-oncology

13. Sarcoma vs. Carcinoma: Differences and Similarities – Verywell Health. www.verywellhealth.com/sarcoma-vs-carcinoma-4694486

14. AN Bashkatov, VP Zakharov, AB Bucharskaya, et al. "Malignant tissue optical properties". In *Multimodal Optical Diagnostics of Cancer*, VV Tuchin, J Popp, V Zakharov, Eds. Cham: Springer (2020).

15. GM Cooper. *The cell: A molecular approach*, 2nd ed. Sunderland, MA: Sinauer Associates (2000). The Development and Causes of Cancer. www.ncbi.nlm.nih.gov/books/NBK9963/

16. L Hinck, I Näthke. "Changes in cell and tissue organization in cancer of the breast and colon". *Curr Opin Cell Biol* 26: 87–95 (2014).

17. www.nature.com/scitable/topicpage/cell-division-and-cancer-14046590/

18. R Lugano, M Ramachandran, A Dimberg. "Tumor angiogenesis: causes, consequences, challenges and opportunities". *Cell Mol Life Sci* 77(9): 1745–1770 (2020).

19. C Yao, S Wu, J Kong, et al. "Angiogenesis in hepatocellular carcinoma: mechanisms and anti-angiogenic therapies". *Cancer Biol Med* 20(1): 25–43 (2023).

Appendix B: Light Propagation in Tissues

The light propagation in tissues can be described by a transport equation known as the Boltzmann equation in statistical mechanics.

However, this integro-differential equation cannot be solved in a general case. Instead, various approximations are used. In this appendix, we present several models of light propagation in tissues referred to in the book's main body.

The solution to the transport equation depends significantly on the light source. In medicine, two types of light sources are commonly used: collimated and diffuse illumination. As they are described by different processes, it is beneficial to consider them separately.

If the collimated light travels through highly absorbing media with small scattering (absorption dominating regime), then its intensity with distance will follow the Beer-Lambert law:

$$I_c = I_{c,0} \exp\left(-\mu_\alpha z\right) \tag{B.1}$$

However, in realistic turbid media, the scattering is significant and often dominates over absorption (scattering dominating regime). In this case, the collimated light experiences scattering and absorption events. As a result, the intensity of the collimated light follows a modified Beer-Lambert law as it propagates within the tissue:

$$I_c = I_{c,0} \exp\left(-\mu_t z\right) \tag{B.2}$$

Here, $\mu_t = \mu_\alpha + \mu_s$ is the total attenuation coefficient.

As a result of light propagation, while some photons are absorbed, some are scattered. As such, collimated light is being converted gradually into diffuse light.

Section B.1 presents a Single Backward Scattering model relevant to certain imaging techniques like OCT.

Then, we will switch to diffuse light propagation in tissues. Several models have been developed to describe the propagation of diffuse light. We will consider two of the most used approximations: the Kubelka-Munk model (section B.2) and diffuse approximation (section B.3). These models assume homogeneous tissue. Section B.4 will demonstrate an approach to extending diffuse approximation to nonhomogeneous tissue.

B.1 SINGLE BACKWARD SCATTERING MODEL

We will develop a simple model that will help us derive metrics, including sampling depth. In doing so, we will follow the Single Backward Scattering model developed in [1].

B.1.1 Individual Scattering Event

Individual scattering events in body tissues are primarily attributed to intracellular organelles (mitochondria, nuclei, etc.) with a size comparable to the photon wavelength (Mie regime). These scattering events are highly anisotropic and primarily directed forward. We will be interested in the probability of backward (with scattering angle $\theta > \pi/2$) scattering p. This probability can be calculated directly from the Mie theory. However, statistically, the angular distribution of scattering is well described by the Heiney-Greenstein scattering phase function [2], which is routinely used for light propagation in biotissues:

$$p(g,\theta) = \frac{1}{2}\frac{1-g^2}{\left(1+g^2-2g\,cos\theta\right)^{3/2}} \tag{B.3}$$

Here, g is the anisotropy coefficient, which is "the mean cosine" of scattering, $<cos\theta>$. If $g = 0$, we have isotropic scattering; $g = 1$ or -1 corresponds to fully forward or fully backward scattering, respectively entirely. The light scattering in tissues is strongly anisotropic, with g ranging from 0.785 (dermis) to 0.995 (blood).

From Eq.B.3, we can estimate the probability of backward scattering p – the probability that photon will be scattered with $\theta > \pi/2$: $p(g) = \int_{\pi/2}^{\pi} p(g,\theta)\,sin\theta d\theta$

$$p(g) = \frac{1}{2}\frac{1-g^2}{g}\left[\frac{1}{\sqrt{1+g^2}} - \frac{1}{1+g}\right] \tag{B.4}$$

It can be seen that the probability of backward scattering for body tissues p is not bigger than 0.05 (dermis), and the typical value of p in tissues with blood is even smaller.

B.1.2 Multiple Scattering in Turbid Tissues

Let us consider the path of a photon that started and finished on the tissue surface (reflection geometry). We can classify the optical paths based on the number of backscattering events (as defined in the previous section) along this path. Optical paths obtained by Monte Carlo simulations can be visually grouped into three major classes: (1) quasi-1D paths with one or more backward scatterings, (2) "sausage" or "banana shape" paths with multiple tilted forward scatterings, and (3) pure diffuse paths with multiple forward and backward scattering events. We can distinguish between paths with one backward scattering and multiple (more than one) backward scatterings. Still, we also can consider paths with tilted forward scattering as a subset of the diffuse paths. Thus, the amended classification can be presented as:

1) Path with one backward scattering

2) Path with multiple backward scatterings

3) Diffuse path, including path with tilted forward scatterings

Let's consider the contributions of each of these scenarios in the total reflected flux. To calculate this contribution, we consider a homogeneous medium with coefficients of absorption μ_a and scattering μ_s, respectively. We also consider the external light flux's normal (or almost normal) incidence to simplify our calculations.

B.1.2.1. Path with one Backward Scattering

Let's consider that each scattering event occurs on the distance $l_s = 1/\mu_s$ (this assumption can be relaxed; however, it will complicate the calculations). Such as forward and almost forward scattering events dominate substantially, so can we consider the following quasi-1D model (the path is almost ballistic; however, it can slightly deviate in transverse directions): if the photon was scattered n times forwardly (with probability $1 - p$) and then was scattered backwardly (with probability p), then to reach the surface again it must undergo additionally n forward scattering events. The probability of this $n - 1 - n$ path is $p(1 - p)^{2n}$. However, we have not considered that the photon flux also undergoes absorption during this path.

To take absorption into account, let us consider the photon flux propagation between two scattering events. The intensity of outcoming flux $I_{out} = I_{in} exp(-\mu_a l_s) = I_{in} exp(-\mu_a / \mu_s)$. So, we can attribute an absorption multiplier (propagator) $a = exp(-\mu_a / \mu_s)$ to each pathway between the two closest scattering events.

Thus, in the previously considered case of $n - 1 - n$ scatterings, we have to multiply scattering probability $p(1 - p)^{2n}$ on the probability of surviving in absorption between $2n + 1$ scattering events: a^{2n+2}. So, the total contribution of this path to the total backward flux will be $p(1 - p)^{2n} a^{2n+2}$.

Thus, the total contribution of all paths with single backward scattering (the coefficient of reflectance in a single backward scattering model) will be provided by the infinite series:

$$R_1 = pa^2 + p(1-p)^2 a^4 .... + p(1-p)^{2n} a^{2n+2} ....$$ (B.5)

It is easy to calculate this sum if we take into account that it is a geometric series. As $(1-p)^2 a^2 < 1$, the series converges and its sum is equal to

$$R_1 = \frac{pa^2}{1-(1-p)^2 a^2}$$ (B.6)

We can also develop a continuous version of our approach. For this reason, we in Eq.B.5 switch from summation to integration: $\sum_{n=0}^{\infty} \to \int dn$. Thus, if we consider a semi-finite tissue, then we can integrate from zero to infinity, and our integral is equal to

$$R_1 = \frac{pa^2}{2\left(\dfrac{\mu_a}{\mu_s} - \ln(1-p)\right)}$$ (B.7)

B.1.2.2 Path with Multiple Backward Scattering

It is easy to see that if we exclude purely diffuse paths and limit ourselves to a quasi-1D quasi-ballistic model, then the total number of backward

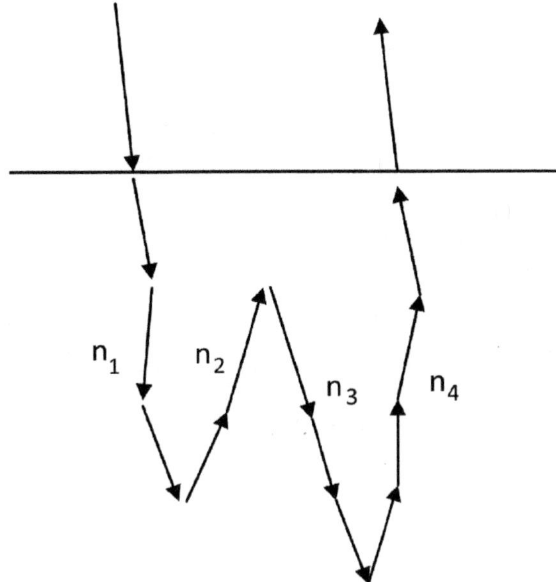

FIGURE B.1　Threefold backward scattering.

scattering events must be odd. So, our next diagram will be one with three backward scattering events (see Figure B.1).

These lowest-order diagrams will be proportional to $p^3 a^6$. To each diagram with the total path length $2N$ we can assign a weight $W(2N,3)$ to the number of different diagrams with the same length of $2N$ and three backward scatterings. This weight can be estimated from the geometry of this path. For our n_1-1-n_2-1-n_3-1-n_4 scattering path (here, n_i is the i-th path length between two backward scattering events), we can write: $n_1 + n_3 = n_2 + n_4 = N$. Thus, we have $(N\text{-}2)^2$ nonequivalent diagrams (such as $1 < n_1,\ n_4 < N$). And the contribution of these diagrams can be approximated as $(N-2)^2 p^3 (1-p)^{2N-4} a^{2N}$. It is easy to see that for strongly anisotropic body tissues with blood we can neglect them in the future. A more accurate estimation of multi-scattering contribution is provided in [1]. However, given that $p \ll 1$, we can expect that for $\mu_a / \mu_s > 0.01$ the impact of all p^{2l+1} where $l > 1$ is relatively small. Multiple backward scattering can have a measurable impact for small g (less than 0.5, which does not represent significant interest for biological tissues). However, the validity of the quasi-1D model for $g < 0.5$ is very questionable, and this impact will likely be masked by the contribution of diffuse paths.

B.1.2.3 Diffuse Path

The contribution of diffuse paths is well known in diffusion theory and can be calculated using various approaches. We can estimate reflectance near the light entry point by visualizing light propagation as a random walk on a cubic lattice with the unit size $l'_s = 1/\mu'_s$ [3]. Thus, if the light enters the tissue at point (x,y,0), it travels to (x,y,1), and from that point occurs random walk through the media. However, to exit tissue near the entry point, it should travel again through (x,y,1) to (x,y,0). On each step flux occurs attenuation $\exp\left(-\left(\mu_a + \mu'_s\right)l'_s\right) = \exp(-1/a')$, where $a' = \mu'_s / \left(\mu_a + \mu'_s\right)$ is the reduced albedo. So, the reflectance at the light entry point (x,y,0) in the lattice model will be $R_d = \exp\left(-\dfrac{2}{a'}\right)\left(\dfrac{1}{6} + C\right)$,

where one-sixth is attributed to the shortest path ((x,y,0)->(x,y,1)->(x,y,0)) and C is the contribution of all paths with nonzero length, which started and finished at (x,y,1). It is easy to see that the photon has to undergo $2/(1 - g)$ scattering events to reach the surface. Thus, the impact of diffuse paths can be quite low in the case of small reduced albedo (e.g., $a' < 0.67$ if the absorption is high or g is close to 1 in case of moderate absorption).

Thus, we have found that single backward scattering can dominate among scattering mechanisms in turbid tissues under certain conditions, such as high absorption.

B.2 KUBELKA-MUNK MODEL

Kubelka and Munk [4] developed an elegant light transport model in certain geometries, such as a layer of paint. Initially, the model was developed for the paper industry. However, due to its simplicity and intuitiveness, occasionally it is used in tissue optics.

The Kubelka-Munk (KM) model considers a slab of homogenous material with some finite thickness L but infinite width and length (infinite sheet approximation), illuminated by ambient (diffuse) light. Based on this geometry, the light propagation is considered one-dimensional. Two hemispherical fluxes model light propagation: forward flux I and backward flux J, which propagate in opposite directions and experience absorption with coefficient K [1/cm] and scattering with coefficient S [1/cm]. In this case, the light propagation can be described by two differential equations:

$$\frac{dI}{dz} = -\left(K + S\right)I + SJ \tag{B.8}$$

$$\frac{dJ}{dz} = -(K+S)J + SI \tag{B.9}$$

This system of equations can be solved for arbitrary thickness L. However, boundary conditions need to be taken into account. For example, we can consider a thin layer placed on a substrate with reflectance R_g. In this case, the reflectance R and transmittance T of the slab can be found:

$$R = \frac{1 - R_g \left[a - b \times coth(bSL)\right]}{a - R_g + b \times coth(bSL)} \tag{B.10}$$

$$T = \frac{b}{a \times sinh(bSL) + b \times cosh(bSL)} \tag{B.11}$$

where $a = 1 + K/S$, and $b = \sqrt{a^2 - 1}$.

However, in most cases, the tissue can be approximated as a semi-infinite. In this case, the solution is straightforward

$$R = a - b = 1 + \frac{K}{S} - \sqrt{(1 + K/S)^2 - 1} \tag{B.12}$$

By construction, the KM model considers only backward scattering, so $g = -1$.

KM coefficients K and S are related to traditional absorption μa and scattering μs coefficients used in tissue optics. In particular, for completely diffused light, Kubelka [5] suggested that $K = 2\mu_a$ and $S = 2\mu_s$. However, other approaches are typically used. For example, van Gemert and Star [6] obtained another expression for scattering: $S = 3/4\mu_s(1-g) - 1/4\mu_a$.

Note that for the one-dimensional approach to be valid, the lateral dimensions need to be much larger than the mean free paths, $1/K$ and $1/S$, for absorption and scattering in the material.

The theory can be extended to more complex cases. For example, it can accommodate the presence of mismatched boundaries or multilayerness.

B.3 DIFFUSE APPROXIMATION

The light traveling in tissue experiences multiple scattering events. Thus, gradually, it forgets its initial direction. At some level, the light transport in tissue can be considered a random walk. Consequently, the light transport at this level can be described by diffusion equation, which becomes a Laplace equation in the steady-state case. The diffuse approximation is a realistic model to describe light propagation in a scattering-dominating regime ($\mu_a \ll \mu_s$).

The characteristic scale of diffusion approximation is $\dfrac{1}{\mu_s'} = \dfrac{1}{\mu_s(1-g)}$. The photon forgets its initial direction at this distance, and the "walk" becomes random.

Thus, in diffuse approximation, the tissue optical parameters are fully characterized by two parameters: absorption coefficient μ_a and reduced scattering coefficient $\mu_s' = \mu_s(1\text{-}g)$.

The theory of diffuse approximation is well described elsewhere (see, for example, [7]). Here, we will present several results relevant to imaging geometries.

For a semi-infinite medium with *wide beam diffuse illumination*, the total radiant energy fluence rate within the tissue far from the borders of the beam depends only on the depth z (Eq. 6.88 in [7]):

$$\varphi_d(z) = \frac{4}{1-r_{10}} \frac{\exp\left(-\mu_{eff}z\right)}{1+h\mu_{eff}} \tag{B.13}$$

Where $\mu_{eff} = \sqrt{\mu_a/\delta}$, $\delta = 1/3\mu_{tr}$, $\mu_{tr} = \mu_a + \mu_s(1-g)$, r_{10} – is the coefficient of reflection of diffuse light on the border of tissue and air (*r10* can be approximated using the relative index of refraction n: $r_{10} \approx 1\text{-}n^{-2}$), $h = 2\delta\dfrac{1+r_{10}}{1-r_{10}}$, μ_a, μ_s, and g are coefficients of absorption, scattering, and anisotropy. This expression needs to be multiplied by the surface density of the incident light, which we assumed 1 (W/m²) in Eq.B.13 and B.14.

Similarly, we can solve the semi-infinite problem for *wide beam collimated illumination*. The difference here is the presence of a collimated term, which dissipates proportionally to $\exp\left(-(\mu_a+\mu_s)z\right)$ (See Eq.B.2). For the biologically relevant case, $\mu_a \ll \mu_s$, we have an expression (Eq. 6.83 in [7]):

$$\varphi_c(z) = \frac{5-r_{10}}{1-r_{10}} \frac{\exp\left(-\mu_{eff}z\right)}{1+h\mu_{eff}} - 2\exp\left(-(\mu_a+\mu_s)z\right) \tag{B.14}$$

Note that the intensity of diffuse light decays according to $\exp\left(-\mu_{eff} z\right)$ law. The coefficient $\mu_{eff} = \sqrt{3\mu_a\left(\mu_a + \mu_s\left(1-g\right)\right)}$ is known as the effective attenuation coefficient.

B.4 PERTURBATIONS IN DIFFUSE APPROXIMATION

While we can estimate penetration depth, mean optical path, and sampling depth, this information does not directly answer whether a certain object (e.g., blood vessel) will be visible and under which conditions. To answer these questions directly, we will develop another analytical approach. In doing so, we will follow [8, 9].

We can consider the following model.[1] The homogeneous semi-infinite tissue is characterized by an absorption coefficient μ_a, and "defect" is described by the volume V and absorption coefficient $\mu_a + \delta\mu_a$ and located at the depth Z (here $\delta\mu_a$ is an incremental absorption coefficient associated with the defect).

B.4.1 Mathematical Model

The light propagation problem in homogeneous tissue can be solved exactly for specific geometries (e.g., slab, semi-space, or spheroid in diffuse approximation, slab, and semi-space in the Kubelka-Munk model (see, e.g. [1]). However, an arbitrary defect significantly complicates things, and we must look for an approximate solution. A perturbation theory can be a helpful approach to finding such an approximate solution: we start from the exact solution for the semi-space geometry and add the defect as a perturbation. Our perturbation approach consists of the following steps:

1. We solve the light propagation problem in homogeneous semi-infinite tissue (the radiant energy fluence rate $\varphi\left(\vec{\rho}\right)$ (W/m²)).

2. Suppose we know the radiant energy fluence rate $\varphi\left(\vec{\rho}\right)$ at some particular point $\vec{\rho}$. We can calculate additional (or incremental) absorbed optical power density $\delta\mu_a\varphi\left(\vec{\rho}\right)$ (W/m³) for some optical heterogeneity with the absorption coefficient $\mu_a + \delta\mu_a$ located at this point. If the volume of the heterogeneity is V, then the additional power absorbed at this heterogeneity will be $\delta\mu_a\varphi\left(\vec{\rho}\right)V$ (W).

3. Alternatively, the heterogeneity can be considered a negative (or inverse) point source with power $-\delta\mu_a V\varphi(\vec{\rho})$ located at the point $\vec{\rho}$. The radiant energy fluence rate induced by such a source can be calculated exactly.

This problem can be analyzed using the diffuse approximation.

Step 1: We can solve the light propagation problem in homogeneous semi-infinite tissue (the radiant energy fluence rate $\varphi(\vec{\rho})$) for wide beam diffuse illumination (see Eq.B.13) and for wide beam collimated illumination (see Eq.B.14).

Step 2: The additional power absorbed at the inhomogeneity can be found by multiplication of Eq.B.13 on $V\delta\mu_a$.

Step 3: The diffuse source with power P in isotropic mediums generates radiant energy fluence rate on the distance ρ from the source.

$$\varphi_s(\rho) = \frac{3P\mu_{tr}}{4\pi\rho}\exp\left(-\mu_{\mathit{eff}}\rho\right) \tag{B.15}$$

Thus, we can represent our defect as the point source described by Eq.B.15, where power P is calculated in step 2 with a minus sign (negative source). To consider the boundary conditions, we can use the diffusion dipole model [10, 11], and in addition to the initial source located at depth Z, consider the second source (with opposite sign) located on the distance $2h + Z$ above the surface. In this case, total flux approximately satisfies realistic boundary conditions for all r ($r = \sqrt{x^2 + y^2}$) and $z=0$ [12]

$$\varphi(r,z) - h\frac{\partial\varphi(r,z)}{\partial z} = 0 \tag{B.16}$$

B.4.2 Contrast Ratio

We can use a contrast ratio defined by Eq.3.4. In our model, the flux on the surface of the tissue consists of two parts $\varphi(x,y) = \varphi + \varphi_s(x,y)$, where φ is a nunperturbed flux from a homogeneous tissue (does not depend on $<x,y>$) and $\varphi_s(x,y)$ is a flux caused by a negative point source (defect), respectively. We can take the unperturbed flux on the surface as a background ($\varphi_b = \varphi$). We cannot ignore the flux from the defect, $\varphi_s(x, y)$ near the inhomogeneity. Thus, we can write:

$$c(x, y) = \frac{\varphi_b - \varphi(x, y)}{\varphi_b} = -\frac{\varphi_s(x, y)}{\varphi_b} \tag{B.17}$$

B.4.3 Results

If the point source with power P is located at $(0, 0, Z)$, then the fluence rate at any point on the tissue surface (here we assume cylindrical coordinates) in the presence of mismatched boundary (Eq.B.16) will be:

$$\varphi_s(r) = \frac{3P\mu_{tr}}{4\pi} \left[\frac{\exp\left(-\mu_{eff}\left(Z^2 + r^2\right)^{1/2}\right)}{\left(Z^2 + r^2\right)^{1/2}} - \frac{\exp\left(-\mu_{eff}\left((2h + Z)^2 + r^2\right)^{1/2}\right)}{\left((2h + Z)^2 + r^2\right)^{1/2}} \right] \tag{B.18}$$

Here again $\mu_{eff} = \sqrt{\mu_a / \delta}$, $\delta = 1/3\mu_{tr}$, $\mu_{tr} = \mu_a + \mu_s(1 - g)$, $h = 2\delta\dfrac{1 + r_{10}}{1 - r_{10}}$

where r_{10} – is the coefficient of reflection of diffuse light on the border of tissue and air. As previously defined, r is the distance on the surface of the tissue from the projection of the defect to the surface ($r = \sqrt{x^2 + y^2}$).

As we already discussed, the power of the source P can be assessed as $-\delta\mu_a\phi(Z)V$, where $\phi(Z)$ can be found from Eq.B.13 (diffuse illumination) or Eq.B.14 (collimated illumination).

B.4.3.1 Single-Point Defect

The advantage of the wide-beam diffuse illumination scenario is that it allows for obtaining closed-form expressions. Diffuse illumination is a quite realistic scenario (e.g., ambient light); in this section we will limit ourselves to this scenario only.

We will keep considering the inhomogeneity located at $(0, 0, Z)$ and using cylindrical coordinates. Far from the inhomogeneity, its effect on flux on the surface is negligible. Thus, we can take the flux rate on the surface at this point as a background ($\phi_b = \phi_d(0)$). We cannot ignore the flux rate from the inhomogeneity, $\phi_s(r)$, near the defect. If we compare the background flux with the fluence rate on the surface in the presence of the inhomogeneity ($\phi(r) = \phi_d(0) + \phi_s(r)$), we can calculate the contrast ratio at any point on the surface of the tissue:

$$c(r) = \frac{\varphi_b - \varphi(r)}{\varphi_b}$$

$$= \frac{3\mu_{tr}\delta\mu_a V exp\left(-\mu_{eff}Z\right)}{4\pi}\left[\frac{exp\left(-\mu_{eff}\left(Z^2 + r^2\right)^{1/2}\right)}{\left(Z^2 + r^2\right)^{1/2}} - \frac{exp\left(-\mu_{eff}\left((2h+Z)^2 + r^2\right)^{1/2}\right)}{\left((2h+Z)^2 + r^2\right)^{1/2}}\right]$$

$$\text{(B.19)}$$

Immediately above the heterogeneity *(r=0)* from Eq.B.19, we can get a compact expression:

$$c(0) == \frac{3\mu_{tr}\delta\mu_a V exp\left(-2\mu_{eff}Z\right)}{4\pi}\left[\frac{1}{Z} - \frac{exp\left(-\mu_{eff}2h\right)}{2h+Z}\right] \qquad \text{(B.20)}$$

B.4.3.2 Linear Defects

The approach can be expanded to linear defects arranged horizontally or vertically. The details of the calculations are presented in [9]. Here, we present only results.

Single-point, horizontally and vertically arranged defects can be evaluated numerically. For these purposes, we split the single point defect volume V as $V = dXdYdZ$ and assumed that the $dX = dY = dZ = 20$ μm. In the case of horizontally and vertically arranged defects, we integrated over x and z, respectively, keeping the same cross-section area ($S = 20$ μm $\times$ 20 μm).

If we take realistic assumptions for mucosa: $\alpha_a = 0.001$ mm^{-1}, $\mu'_{s'} = 5$ mm^{-1} (at 532 nm) [13], $\delta\mu_a = 28$ mm^{-1} (the whole blood with 70% oxygenation at 532 nm), and $n = 1.33$, we can calculate the contrast ratio for various defect geometries.

While for the single point and the horizontally arranged defects, the outcome depends on the volume (or cross-section) and the defect depth, Z only, the vertically arranged defect outcome depends on the defect height,

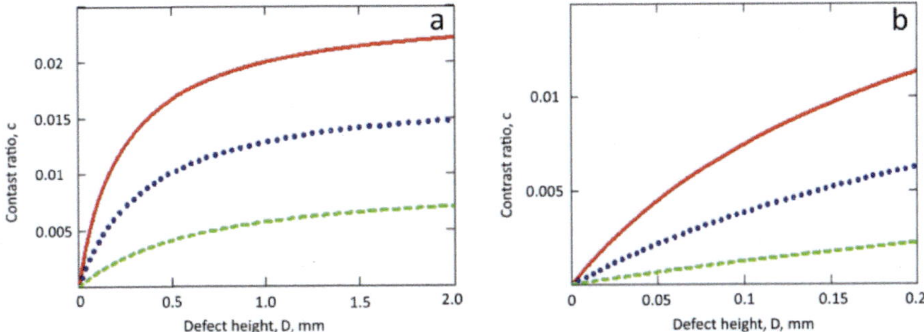

FIGURE B.2 Vertically arranged defects. The dependence of the contrast ratio above the defect ($r = 0$) as a function of the defect height, D, for $Z = 0.1$ mm (red solid line), 0.2 mm (blue dotted line), and 0.5 mm (green dashed line). Panel a: full physiological range (dermis, 2 mm). Panel b: zoomed-in area (papillary dermis, 0.2 mm).

D. Thus, to compare all three scenarios, firstly, we need to understand that dependence.

In Figure B.2, one can see the dependence of the maximal contrast ratio (just above the defect, $r = 0$) in the case of the vertical defect as a function of the defect height, D. We evaluated it for the entire thickness of the dermis layer (2 mm, panel a) and the reasonable estimation of the nurturing capillary height (0.2 mm, panel b).

In Figure B.3, one can see a comparison of the dependence of the contrast ratio on the distance for the point defect (left pane), horizontally, and vertically arranged defects (central and right pane, respectively). Defect depth was $Z = 0.1$ mm (solid red line), 0.2 mm (blue dotted line), and 0.5 mm (green dashed line). In the case of a vertical defect, the defect height was set at $D = 0.2$ mm (nurturing capillaries).

B.4.3.3 Illumination

We can also analyze how the contrast will differ for the same defect in the case of diffuse and collimated illumination. We will do it for a single-point defect. As discussed in section B.4.2, the contrast at any point on the surface of the tissue for the defect located on the depth Z would be $c(r) = -\varphi_s(r) / \varphi(0)$ (see Eq.B.17). Here $\varphi(z)$ is the unperturbed flux distribution for either diffuse light (Eq.B.13) or collimated light (Eq.B.14). Thus, taking into account that for our negative source $\varphi_s(r) \sim -\delta\mu_a V \varphi(Z) \sim \varphi(Z)$

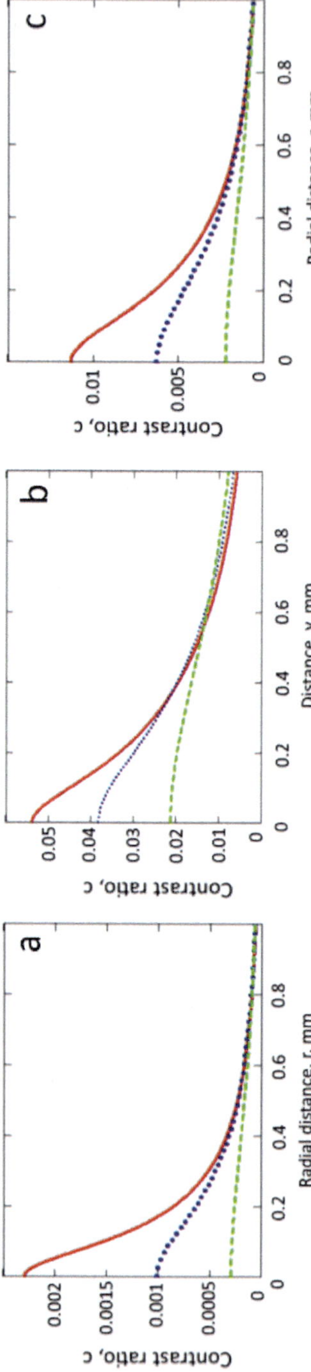

FIGURE B.3 The radial dependence of the contrast ratio for the point defect (panel a), horizontally and vertically arranged defects (panels b and c, respectively). Defect depth, $Z = 0.1$ mm (red solid line), 0.2 mm (blue dotted line), and 0.5 mm (green dashed line). In the case of a vertical defect, the defect height, $D = 0.2$ mm.

after simple reducing we can find that the ratio of contrasts for collimated light and diffuse light will be

$$\frac{c_c}{c_d} = \frac{\varphi_c(Z)/\varphi_c(0)}{\varphi_d(Z)/\varphi_d(0)} = \frac{a}{a-2} - \frac{2}{a-2}\exp\left(-\left(\mu_a + \mu_s - \mu_{eff}\right)Z\right) \quad \text{(B.21)}$$

Here, for the sake of brevity, we introduced $a = \dfrac{5-r_{10}}{1-r_{10}}\dfrac{1}{1+h\mu_{eff}}$.

We have analyzed the problem in a realistic case of the upper part of the capillary loop in the dermis: $\mu_a = 0.033$ mm^{-1} and $\mu_s' = 5$ mm^{-1} for reticular dermis [13], $\delta\mu_a = 28$ mm^{-1} (the whole blood with 70% oxygenation at 532 nm), $V = 2 \times 10^3$ μm^3, $n = 1.33$. Results are presented in Figure 4.4.

From Figure 4.4B, one can see that such defects can be visualized (with $c_{th} = 0.001$) till approximately 0.059 mm for diffuse illumination and 0.078 mm for collimated illumination.

NOTE

1 For consistency, we will use capital letters to indicate defect properties. E.g., V – defect volume, Z – defect depth, D – defect height. However, we will use the traditional notation for the defect's optical properties (e.g., $\delta\mu_a$ is an incremental absorption coefficient associated with the defect). A point in the bulk of the tissue (3D) and on the surface of the tissue (2D) will be denoted (x,y,z) and <x,y>, respectively. We will also use "defect" and "inhomogeneity" interchangeably.

REFERENCES

1. G Saiko, A Douplik. "Reflectance of biological turbid tissues under wide area illumination: Single backward scattering approach". *Int J Photoenergy* 241364 (2014).
2. LG Henyey, JL Greenstein. "Diffuse radiation in the galaxy". *Nature* 147(3733): 613–626 (1941).
3. G Saiko, A Douplik. "Real-time optical monitoring of capillary grid spatial pattern in epithelium by spatially resolved diffuse reflectance probe". *J Innov Opt Health Sci* 5(2) (2012).
4. P Kubelka, F Munk. "Ein Beitrag zur Optik der Farbanstriche". *Z Tech Phys (Leipzig)* 12: 593–601 (1931).
5. P Kubelka. "New contributions to the optics intensely lightscattering materials. Part I". *J Opt Soc Am* 38: 448–457 (1948).
6. MJC van Gemert, WM Star. "Relations between the Kubelka- Munk and the transport equation models for anisotropic scattering". *Lasers Life Sci* 1: 287–298 (1987).

7. WM Star. "Diffusion theory of light transport". In *Optical-Thermal Response of Laser-Irradiated Tissue*, 2nd ed., AJ Welch, Ed. Dordrech, NLD: Springer, p. 178 (2011).

8. G Saiko, A Douplik. "Contrast ratio during visualization of subsurface optical inhomogeneities in turbid tissues: Perturbation analysis". In *Proceedings of the 14th International Joint Conference on Biomedical Engineering Systems and Technologies (BIOSTEC 2021) – Vol. 2: BIOIMAGING*, pp. 94–102 (2021).

9. G Saiko, A Douplik. "Visibility of capillaries in turbid tissues: An analytical approach". *arXiv*: 2210.04301 (2022).

10. RJ Frerrerd, RL Longini. "Diffusion dipole source". *J Opt Soc Am* 63: 336–337 (1973).

11. A Kienle, MS Patterson. "Improved solutions of the steady-state and the time-resolved diffusion equations for reflectance from a semi-infinite turbid medium". *J Opt Soc Am A* 14: 246–254 (1994).

12. RC Haskell, LO Svaasand, T-T Tsay, et al. "Boundary conditions for the diffuse equation in radiative transfer". *J Opt Soc Am A* 11: 2727–2741 (1994).

13. IV Meglinski, SJ Matcher. "Quantitative assessment of skin layers absorption and skin reflectance spectra simulation in the visible and near-infrared spectral regions". *Physiol Meas* 23: 741753 (2002).

Index

Note: Page numbers in *italics* indicate a figure on the corresponding page.